Advanced Studies of
Thermoelectric Systems

Advanced Studies of Thermoelectric Systems

Diana Enescu

Basel • Beijing • Wuhan • Barcelona • Belgrade • Novi Sad • Cluj • Manchester

Diana Enescu
Department of Electronics,
Telecommunications and Energy
Valahia University of Targoviste
Targoviste
Romania

Editorial Office
MDPI AG
Grosspeteranlage 5
4052 Basel, Switzerland

This is a reprint of articles from the Special Issue published online in the open access journal *Energies* (ISSN 1996-1073) (available at: www.mdpi.com/journal/energies/special_issues/ Advanced_Studies_of_Thermoelectric_Systems).

For citation purposes, cite each article independently as indicated on the article page online and using the guide below:

Lastname, A.A.; Lastname, B.B. Article Title. *Journal Name* **Year**, *Volume Number*, Page Range.

ISBN 978-3-7258-2056-6 (Hbk)
ISBN 978-3-7258-2055-9 (PDF)
https://doi.org/10.3390/books978-3-7258-2055-9

© 2024 by the authors. Articles in this book are Open Access and distributed under the Creative Commons Attribution (CC BY) license. The book as a whole is distributed by MDPI under the terms and conditions of the Creative Commons Attribution-NonCommercial-NoDerivs (CC BY-NC-ND) license (https://creativecommons.org/licenses/by-nc-nd/4.0/).

Contents

About the Editor vii

Preface ix

Diana Enescu
Innovations in Thermoelectric Technology: From Materials to Applications
Reprinted from: *Energies* 2024, 17, 1692, doi:10.3390/en17071692 1

Alexander Vargas-Almeida, Miguel Angel Olivares-Robles and Andres Alfonso Andrade-Vallejo
Design of Thermoelectric Generators and Maximum Electrical Power Using Reduced Variables and Machine Learning Approaches
Reprinted from: *Energies* 2023, 16, 7263, doi:10.3390/en16217263 8

Julia Camut, Eckhard Müller and Johannes de Boor
Analyzing the Performance of Thermoelectric Generators with Inhomogeneous Legs: Coupled Material–Device Modelling for Mg_2X-Based TEG Prototypes
Reprinted from: *Energies* 2023, 16, 3666, doi:10.3390/en16093666 35

Julian Schwab, Christopher Fritscher, Michael Filatov, Martin Kober, Frank Rinderknecht and Tjark Siefkes
Experimental Analysis of the Long-Term Stability of Thermoelectric Generators under Thermal Cycling in Air and Argon Atmosphere
Reprinted from: *Energies* 2023, 16, 4145, doi:10.3390/en16104145 53

Wenlong Yang, Wenchao Zhu, Yang Yang, Liang Huang, Ying Shi and Changjun Xie
Thermoelectric Performance Evaluation and Optimization in a Concentric Annular Thermoelectric Generator under Different Cooling Methods
Reprinted from: *Energies* 2022, 15, 2231, doi:10.3390/en15062231 63

Alessandro Bellucci, Stefano Orlando, Luca Medici, Antonio Lettino, Alessio Mezzi and Saulius Kaciulis et al.
Nanostructured Thermoelectric PbTe Thin Films with Ag Addition Deposited by Femtosecond Pulsed Laser Ablation
Reprinted from: *Energies* 2023, 16, 3216, doi:10.3390/en16073216 84

Majed Alshammari, Turki Alotaibi, Moteb Alotaibi and Ali K. Ismael
Influence of Charge Transfer on Thermoelectric Properties of Endohedral Metallofullerene (EMF) Complexes
Reprinted from: *Energies* 2023, 16, 4342, doi:10.3390/en16114342 98

Diana Enescu
Heat Transfer Mechanisms and Contributions of Wearable Thermoelectrics to Personal Thermal Management
Reprinted from: *Energies* 2024, 17, 285, doi:10.3390/en17020285 107

Anna Dabrowska, Monika Kobus, Łukasz Starzak and Bartosz Pekosławski
Evaluation of Performance and Power Consumption of a Thermoelectric Module-Based Personal Cooling System—A Case Study
Reprinted from: *Energies* 2023, 16, 4699, doi:10.3390/en16124699 136

Stephen Lucas, Romeo Marian, Michael Lucas, Titilayo Ogunwa and Javaan Chahl
Employing the Peltier Effect to Control Motor Operating Temperatures
Reprinted from: *Energies* 2023, 16, 2498, doi:10.3390/en16052498 152

Mohammad Tariq Nasir, Diaa Afaneh and Salah Abdallah
Design Modifications for a Thermoelectric Distiller with Feedback Control
Reprinted from: *Energies* **2022**, *15*, 9612, doi:10.3390/en15249612 . . . . **167**

About the Editor

Diana Enescu

Dr. Diana Enescu is currently a researcher at the National Metrology Institute (INRiM), Torino, Italy, and teaches courses in the Electronics, Telecommunications, and Energy Department at Valahia University of Targoviste, Targoviste, Romania. Her research interests include thermal engineering, energy efficiency, and renewable energy technologies, focusing on thermoelectric systems. Dr. Enescu has participated in various international research projects. She is also recognized for her extensive publications in high-impact journals and her role as a guest editor for special issues related to energy systems. Dr. Enescu aims to develop sustainable energy solutions and enhance the practical applications of thermoelectric systems.

Preface

This reprint, "*Advanced Studies of Thermoelectric Systems*", is a collection of research contributions referring to the latest advancements in thermoelectric technology, focusing on the design, optimization, and application of thermoelectric generators (TEGs). The topics of this reprint include both theoretical fundamentals and practical applications, offering valuable insights into the conversion of waste heat into usable electrical energy—a technology that is increasingly vital in diverse fields, from automotive and wearable electronics to hypersonic vehicles and energy harvesting.

The motivation behind this work arises from the rapid progress in thermoelectric research, followed by the need for more efficient and sustainable energy solutions. This reprint addresses key challenges such as improving material properties, reducing heat loss, and enhancing the scalability of thermoelectric devices. The scope of this reprint includes a wide range of research themes—ranging from novel material development to the integration of TEGs in advanced technological systems—with the aim of advancing the field toward more effective and widespread use.

This reprint is intended for a broad audience, including researchers, engineers, and industry professionals who are interested in the latest developments in energy efficiency, material science, and the practical applications of thermoelectric engineering.

The contributions included in this volume are authored by a distinguished group of experts from various disciplines, reflecting the collaborative nature of the research presented. The involved authors have brought together their expertise to create a comprehensive overview of the current state and future potential of thermoelectric technology, ensuring the credibility and depth of the content.

I would like to take this opportunity to express my deep gratitude to all the authors for their invaluable contributions. Their expertise and dedication have been crucial in the development of this reprint. Special recognition is also due to the peer reviewers, whose insightful feedback has significantly enhanced the quality of the research presented here. This reprint is a demonstration of the collaborative efforts and shared commitment of everyone involved in making this project happen.

Diana Enescu
Editor

Editorial

Innovations in Thermoelectric Technology: From Materials to Applications

Diana Enescu [1,2]

1. Electronics, Telecommunications and Energy Department, University Valahia of Targoviste, 130004 Targoviste, Romania; diana.enescu@valahia.ro or d.enescu@inrim.it
2. Division of Applied Metrology and Engineering, Istituto Nazionale di Ricerca Metrologica (INRiM), 10135 Torino, Italy

1. Introduction

Over the past two decades, significant advances have been made in the field of energy harvesting, which involves the collection of energy from various environmental sources, including light, thermal gradients, electromagnetic radiation, and mechanical vibrations [1]. Thermal energy is ubiquitous and can be found in electronic devices (such as integrated circuits, smartphones, and computers), vehicles, buildings, and even within the human body, representing a significant and accessible source of energy [2]. Devices known as thermoelectric generators (TEGs) play a crucial role in harvesting this energy by transforming thermal energy directly into electrical energy without the need for mechanical processes, unlike conventional turbine-based power generation methods [3]. TEGs are characterized by energy efficiency, negligible maintenance requirements, and durability, making them increasingly popular in energy harvesting for a broad range of applications. Despite their low efficiency of about 5%, the versatility of TEGs makes them ideal for diverse operational settings, including space use, medical applications within the human body [4], and even aircraft engines [5], where price is not an essential parameter [6]. Small, affordable, and efficient TEG systems are a good alternative to batteries for many uses [7,8]. For example, a TEG system that makes 1.5 mW of power from a temperature difference $\Delta T = 10\ ^\circ C$ can run a small preamplifier and a sensor controller [6]. On the other hand, the Internet of Things (IoT) refers to the network of physical objects embedded with sensors, software, and other technologies for the purpose of connecting and exchanging data with other devices and systems over the Internet. IoT gathers large amounts of data automatically without human aid by using smart, small devices that can sense, compute, and connect to each other. TEG is a good method for converting thermal energy into electricity for these IoT devices, also promoting the use of renewable energy [9]. TEGs are a promising technology for powering wearable electronics and IoT node sensors. They offer a number of advantages over traditional batteries, including their small size, light weight, and ability to harvest energy from a variety of sources [10].

TEGs can be used in vehicles to convert waste heat into electricity. Waste heat energy is described as unconsumed heat energy that is rejected from a thermal process [11]. For example, Kim et al. [12] put a hexagon-shaped TEG in the exhaust pipe of a hybrid electric car. Shu et al. [13] proposed a way to use the waste heat from a diesel engine to run a thermoelectric generator system. This new setup increases its maximum power by 13.4% compared to the old setup. Lan et al. [14] created a model to predict the temperatures and power output of a TEG, which operates using the recovered waste heat energy from a large truck in transit. Li et al. [15] arranged TEG units around the exit of the exhaust pipe of a passenger car. Agudelo et al. [16] explored the capture of waste heat energy from the exhaust gases of a diesel car. Ge et al. [17] also aimed to improve the functioning of this kind of system. Muralidhar et al. [18] conducted a study on the benefits of using a thermoelectric generator in a heavy-duty hybrid electric bus. Their findings indicated that this could lead to reductions of up to 7.2% and 7.58% in fuel use and CO_2 emissions, respectively.

TEGs also offer a promising method for generating power in hypersonic vehicles, which are capable of travelling significantly faster than the speed of sound, usually defined as speeds of at least Mach 5. Hypersonic vehicles are used in a variety of sectors, ranging from military applications, such as missiles and reconnaissance aircraft, to potential future uses in civilian space travel and rapid long-distance transportation. For long-distance travel at hypersonic speeds, creating electricity on the spot is crucial for powering a vehicle, especially in terms of fuel, control, and radar systems [19]. Cheng et al. [20] and Cheng et al. [21] introduced a method for generating power in these vehicles using a sophisticated multi-stage (cascade) TEG system. They managed to reach a peak efficiency of 18.4% by using cutting-edge thermoelectric materials. Furthermore, Cheng et al. [22] conducted a study comparing single-stage and multi-stage TEG setups to identify the most efficient performance conditions for the system.

Recent advancements in thermoelectric materials focus on overcoming the limitations of traditional semiconductors by developing novel materials with higher figures of merit (ZT). The efficiency of thermoelectric materials, crucial for converting heat to electricity, is bound by their Seebeck coefficient, electrical conductivity, and thermal conductivity. While metals offer low electrical resistivity and high thermal conductivity, their low Seebeck coefficients result in poor thermoelectric performance, making semiconductors with high ZT values, such as Bi_2Te_3, PbTe, and $CoSb_3$, more desirable. Recent research has yielded materials such as $Cu_{2-x}Se$ and $PbTe_{0.7}S_{0.3}$, achieving ZT > 2, marking a significant step forward. These innovations, alongside the development of segmented thermocouples designed to operate across different temperature ranges, aim to enhance the efficiency and practicality of thermoelectric generators (TEGs), despite the challenges posed by the high development and industrialization costs of these materials [2].

The above-mentioned aspects indicate the range of current studies on thermoelectric system technologies and applications, and significant progress in different research directions is expected in the coming years, from the evolution of materials to the identification of more effective solutions in the power and energy sector. This context is the motivation behind the collection of specific contributions that form this Special Issue.

2. An Overview of the Contributions to the Special Issue

This Special Issue includes a variety of research papers that provide insights into the technology and applications of thermoelectric systems. The featured papers, written by 42 authors with affiliations in nine countries, cover a range of topics, from theoretical research to practical applications in different fields. This collection not only addresses existing challenges in the field but also opens new areas for future research. The Special Issue is divided into two main themes:

1. TEG design and optimization;
2. Advances and applications of thermoelectric materials and technologies.

For the first theme, TEG design and optimization is a key area for thermoelectric research. Four contributions address specific aspects concerning the geometry, durability, and performance enhancement of TEGs, as follows:

- In "Design of Thermoelectric Generators and Maximum Electrical Power Using Reduced Variables and Machine Learning Approaches", Vargas-Almeida et al. investigated the geometric optimization of TEGs through a combination of a technique to reduce variables and machine learning. The article concludes that combining supervised machine learning with the technique of reduced variables is effective for designing TEGs. This approach is beneficial for estimating thermoelectric properties at various temperatures and designing TEGs with limited amounts of experimental data. The key findings include the ability to analyze and optimize power for different temperature ranges and the potential of using this methodology for TEG design without extensive laboratory resources. Additionally, the study highlights the application of artificial intelligence in predicting thermoelectric properties and electrical power, demonstrating high accuracy and efficiency in modelling TEGs. This study represents a significant advance in predicting thermoelectric properties and maximizing the power output of TEGs, blending computational methods with practical applications.

- In "Experimental Analysis of the Long-Term Stability of Thermoelectric Generators under Thermal Cycling in Air and Argon Atmosphere", Schwab et al. investigated the durability of TEGs under different environmental conditions. The study determined the degradation of TEGs at the system level, particularly emphasizing the impact of the atmosphere on their performance. This degradation is attributed to increased inner resistance, influenced by oxidation and crack formation, especially at electrode interfaces. The experiments demonstrated a significant reduction in the maximum output power in an air atmosphere, which stabilizes at about 57% of its initial maximum power after 50 cycles, while an argon atmosphere mitigates this degradation, maintaining a constant output power during the cyclic tests. The paper recommends developing standard testing procedures based on application-oriented thermal cycles to enhance the design and longevity of TEG systems. The gaps identified in the paper include the need for a broader range of thermoelectric materials and test conditions, as well as a standardized approach to testing, to comprehensively understand and mitigate degradation phenomena in TEGs. This study is crucial for understanding the practical challenges in deploying TEGs in various atmospheric conditions, emphasizing the importance of long-term stability.

- In "Analyzing the Performance of Thermoelectric Generators with Inhomogeneous Legs: Coupled Material–Device Modelling for Mg2X-Based TEG Prototypes", Camut et al. present an innovative method for the performance analysis of TEGs with inhomogeneous material properties. Their study introduces a new analytical approach, the Constant Property for Inhomogeneous Materials (CPIM) approach, which integrates experimental carrier concentration profiling with the Single Parabolic Band (SPB) model and continuum theory to accurately assess the inhomogeneity of thermoelectric materials at macro/mesoscopic levels. Applied to an Mg2(Si,Sn)-based TEG, this method reveals discrepancies between predicted and observed heat flows and thermal losses in N-type legs when considering material inhomogeneity, exceeding measurement uncertainties and suggesting higher thermal losses at TEG/heat exchanger interfaces. The significant difference between calculated and measured electrical resistance points to crack formation as a prevalent degradation mechanism. The CPIM approach, by capturing property inhomogeneity, offers a more accurate analysis of TEG performance under a thermal load, identifying the need for precise heat flow measurements in small TEG prototypes as a critical future challenge. This highlights gaps in current modelling approaches, especially in accounting for material inhomogeneity and accurately measuring thermal and electrical properties in experimental settings.

- In "Thermoelectric Performance Evaluation and Optimization in a Concentric Annular Thermoelectric Generator under Different Cooling Methods", Yang et al. offer a comprehensive analysis of a concentric annular TEG. This study addresses the key aspect of thermal management in TEGs, exploring various cooling methods to optimize performance. This study focused on creating a detailed simulation model that looks into how heat from a car's exhaust can be turned into electricity using a new type of TEG with a special design called a concentric annular heat exchanger. The model considers how temperature affects the materials used in the TEG, how heat is transferred within it, and how the setup of the heat source impacts its efficiency and the temperature differences inside the TEG. The study specifically examined how changing the heat source and cooling methods can affect the TEG's power output and efficiency under various car operating conditions. The study discovered the potential of using advanced TEG systems to recover energy from automotive exhaust fumes, suggesting that with the right cooling method, these systems can significantly improve their efficiency and power output. However, it also highlights gaps, such as the need for more material for optimal performance in some configurations and the varying effectiveness of cooling methods based on the temperature of the heat source.

The second theme includes new developments in thermoelectric materials and their uses in different applications. The contributions on this theme address the use of dedicated materials and discuss the effectiveness of applications relevant to personal comfort and engineering applications. In particular:

- In "Nanostructured Thermoelectric PbTe Thin Films with Ag Addition Deposited by Femtosecond Pulsed Laser Ablation," Bellucci et al. explore the development of nanostructured thermoelectric materials. Their study contributes to the field of nano-materials and their application in thermoelectric devices, focusing on thin films for low-power applications. In this paper, the focus was on enhancing the thermoelectric efficiency of PbTe thin films through the addition of silver (Ag) using Pulsed Laser Deposition (PLD). Previous efforts achieved a lower thermal conductivity in PbTe films by creating nanocrystalline structures, yet these did not significantly improve the power factor compared to bulk materials. Bellucci et al. explored Ag as a dopant, observing its effects on the electrical and thermal properties of the films. Their findings indicate that Ag can reduce resistivity and alter the Seebeck coefficient, suggesting increased carrier concentration. The optimal thermal performance was identified in the 510–540 K range, with the materials showing stability under prolonged heat treatment. Despite achieving an improved power factor with specific Ag concentrations, the performance still fell short of that of bulk materials and other thin film techniques. This gap underscores the need for further exploration into co-doping strategies and complex alloys to advance the thermoelectric properties of thin films.

- In "Influence of Charge Transfer on Thermoelectric Properties of Endohedral Metallo-fullerene (EMF) Complexes", Alshammari et al. analyze the thermoelectric properties of EMFs. The paper focuses on how charge transfer within these complexes can be harnessed to improve electrical conductance and thermopower, offering insights into the material science of thermoelectrics. Their study investigates the charge transfer mechanisms in donor–acceptor complexes, focusing on three different analysis methods: Mulliken population, Hirshfeld, and Voronoi. These techniques were applied to understand how charge is transferred between molecular components. The findings from all three charge transfer methods consistently indicate that charge moves from the metallic components to Ih-C80 cages, with the extent of charge transfer varying depending on the type of metallic moiety involved. This behavior suggests a pathway to enhance both the conductance and thermopower in these complexes, pointing to novel design strategies for electronic and thermoelectric devices by modulating charge transfer through different metallic moieties. However, the study highlights a knowledge gap in comprehensively understanding the relationship between different types of metallic moieties, their specific influence on charge transfer efficiency, and the resultant electronic properties, signaling the need for further targeted research in this area to fully leverage these findings for practical applications.

- In "Heat Transfer Mechanisms and Contributions of Wearable Thermoelectrics to Personal Thermal Management", Enescu discusses the conceptual aspects of heat transfer in the application of wearable TEGs. This paper offers a new viewpoint on wearable TEGs (w-TEGs), focusing on their use for personal thermal management, rather than providing a detailed exploration of the specific materials and devices, as was the case in other reviews. Enescu emphasizes the critical role of understanding how heat transfer between w-TEGs and the human body can enhance local thermal comfort without changing the room temperature. This approach provides innovative ways to improve energy efficiency by considering the body as a heat source and optimizing how the device interacts with the skin through various heat transfer methods, including evaporation. Future progress relies on developing materials with improved thermal performance and designing devices that are flexible, lightweight,

- In "Evaluation of Performance and Power Consumption of a Thermoelectric Module-Based Personal Cooling System—A Case Study," Dąbrowska et al. present an evaluation of a personal cooling system using flexible thermoelectric modules. Their research focused on developing and testing a novel personal cooling system (PCS) for workplace use, using flexible thermoelectric (TE) modules and heat sinks to combat heat effects. The study aimed to assess the effectiveness of these TE modules in reducing thermal discomfort under conditions of increased physical activity and elevated ambient temperatures, while also evaluating the system's electrical power consumption and controller efficiency. Laboratory tests involving human participants compared active cooling by TE modules against passive cooling through wet heat sinks and examined the influence of ambient temperature on the system's performance. The results indicate that TE modules effectively lowered skin temperature over several hours and highlighted the need for the cooling device to adhere closely to the body for optimal heat removal. Additionally, a combination of TE modules with evaporative heat sinks was found to enhance cooling effects. The study also showed that ambient temperatures between 25 °C and 35 °C had a minimal impact on cooling efficiency. A limitation of the study was its reliance on a single participant, suggesting the need for future testing with a broader user base to more comprehensively explore usability, ergonomics, and market viability. This research lays the groundwork for refining the PCS prototype, indicating gaps in understanding user interaction and optimization of TE module count for cost-effectiveness and market success.

- In "Employing the Peltier Effect to Control Motor Operating Temperatures", Lucas et al. examine the application of thermoelectric coolers in electric motors. Their research explores the use of thermoelectric coolers (TECs) not for cooling but for heating electrical motors to minimize thermal gradients and stresses, and to address issues such as internal condensation and water absorption. By reversing the supply polarity, TECs can pump heat energy into the core of the motor, potentially allowing control over internal temperatures, particularly at winding locations. This approach aimed to maintain warmth within the motor's internals and windings. The study involved integrating TECs into three different setups on a standard electrical motor equipped with temperature sensors, marking the first trial of such technology for internal temperature regulation in operational conditions to enhance motor reliability and reduce maintenance costs. The findings revealed that TECs, when mounted on motors using simple or complex adapter plates depending on the motor's exterior design, can significantly increase internal temperatures, surpassing dew points and eliminating condensation problems. This application of TECs presents a cost-effective alternative to traditional heating methods by directly heating the volume of the motor. The potential for TECs to be included in new motor designs during the manufacturing phase was also noted, suggesting a proactive approach to incorporating this technology. While the study successfully demonstrated the capability of TECs to heat motor cores effectively across different configurations, it also opens the door for further investigation into dynamically managing motor temperatures under varying ambient conditions.

- In "Design Modifications for a Thermoelectric Distiller with Feedback Control," Nasir et al. propose a novel design for a thermoelectric distiller. This study introduces a redesigned thermoelectric (TE) distiller that enhances water distillation efficiency by incorporating both the heating and cooling functions of a thermoelectric module.

Key innovations include a new thermoelectric configuration with an added heat sink and cooling fan to better manage the cold tank's temperature, and the development of a feedback control system that optimizes the distiller's performance by adjusting the fan's speed and the TE module's voltage. Specifically, the study explores the use of PID and MPC controllers to maintain the desired operational conditions despite external disturbances and introduces an angled cover design to improve the collection of distilled water. The research assesses the upgraded distiller's effectiveness under various conditions, aiming to maximize pure water production. The findings from the mathematical modelling and comparative analysis of the PID and MPC control systems demonstrate that the MPC controller achieves a superior dynamic performance, enhancing productivity by up to 150% compared to traditional open-loop systems. This productivity boost is attributed to improved heat recovery and vapor production, alongside reduced vapor loss. The advancements outlined in the study suggest a significant step forward in thermoelectric distillation technology, offering a more efficient method for water purification.

3. Concluding Remarks

This Special Issue, entitled "Advanced Studies of Thermoelectric Systems", brings together a collection of innovative papers that highlight the diverse and progressive research on thermoelectric technology. The contributions emphasize how collaboration across various disciplines—such as physics, chemistry, and engineering—leads to significant advances in thermoelectric technology. Such collaboration is fundamental for creating solutions to enhance the energy efficiency of the systems. The importance of the progress made in thermoelectric technology is illustrated by these contributions and highlights directions for advancing the field. Research undertaken in areas such as material science, device design, and optimization strategies addresses major obstacles to the broader application of thermoelectric technology. The major obstacles include the development of materials with improved thermoelectric properties, the reduction in heat loss in devices, and the enhancement of scalability of the practical solutions. Insights derived from these studies offer a deep understanding of the fundamentals of thermoelectric systems and suggest promising directions for ongoing research.

Conflicts of Interest: The author declares no conflicts of interest.

List of Contributions

1. Vargas-Almeida, A.; Olivares-Robles, M.A.; Andrade-Vallejo, A.A. Design of Thermoelectric Generators and Maximum Electrical Power Using Reduced Variables and Machine Learning Approaches. *Energies* **2023**, *16*, 7263. https://doi.org/10.3390/en16217263.
2. Schwab, J.; Fritscher, C.; Filatov, M.; Kober, M.; Rinderknecht, F.; Siefkes, T. Experimental Analysis of the Long-Term Stability of Thermoelectric Generators under Thermal Cycling in Air and Argon Atmosphere. *Energies* **2023**, *16*, 4145. https://doi.org/10.3390/en16104145.
3. Camut, J.; Müller, E.; de Boor, J. Analyzing the Performance of Thermoelectric Generators with Inhomogeneous Legs: Coupled Material–Device Modelling for Mg_2X-Based TEG Prototypes. *Energies* **2023**, *16*, 3666. https://doi.org/10.3390/en16093666.
4. Yang, W.; Zhu, W.; Yang, Y.; Huang, L.; Shi, Y.; Xie, C. Thermoelectric Performance Evaluation and Optimization in a Concentric Annular Thermoelectric Generator under Different Cooling Methods. *Energies* **2022**, *15*, 2231. https://doi.org/10.3390/en15062231
5. Bellucci, A.; Orlando, S.; Medici, L.; Lettino, A.; Mezzi, A.; Kaciulis, S.; Trucchi, D.M. Nanostructured Thermoelectric PbTe Thin Films with Ag Addition Deposited by Femtosecond Pulsed Laser Ablation. *Energies* **2023**, *16*, 3216. https://doi.org/10.3390/en16073216.
6. Alshammari, M.; Alotaibi, T.; Alotaibi, M.; Ismael, A.K. Influence of Charge Transfer on Thermoelectric Properties of Endohedral Metallofullerene (EMF) Complexes. *Energies* **2023**, *16*, 4342. https://doi.org/10.3390/en16114342.
7. Enescu, D. Heat Transfer Mechanisms and Contributions of Wearable Thermoelectrics to Personal Thermal Management. *Energies* **2024**, *17*, 285. https://doi.org/10.3390/en17020285

8. Dąbrowska, A.; Kobus, M.; Starzak, Ł.; Pękosławski, B. Evaluation of Performance and Power Consumption of a Thermoelectric Module-Based Personal Cooling System—A Case Study. *Energies* 2023, *16*, 4699. https://doi.org/10.3390/en16124699.
9. Lucas, S.; Marian, R.; Lucas, M.; Ogunwa, T.; Chahl, J. Employing the Peltier Effect to Control Motor Operating Temperatures. *Energies* 2023, *16*, 2498. https://doi.org/10.3390/en16052498.
10. Nasir, M.T.; Afaneh, D.; Abdallah, S. Design Modifications for a Thermoelectric Distiller with Feedback Control. *Energies* 2022, *15*, 9612. https://doi.org/10.3390/en15249612.

References

1. Vullers, R.J.M.; van Schaijk, R.; Doms, I.; Van Hoof, C.; Mertens, R. Micropower energy harvesting. *Solid-State Electron.* 2009, *53*, 684–693. [CrossRef]
2. Jaziri, N.; Boughamoura, A.; Müller, J.; Mezghani, B.; Tounsi, F.; Ismail, M. A comprehensive review of Thermoelectric Generators: Technologies and common applications. *Energy Rep.* 2020, *6*, 264–287. [CrossRef]
3. He, J.; Li, K.; Jia, L.; Zhu, Y.; Zhang, H.; Jianshe Linghu, J. Advances in the applications of thermoelectric generators. *Appl. Therm. Eng.* 2024, *236*, 121813. [CrossRef]
4. Lossec, M.; Multon, B.; ben Ahmed, H. Sizing optimization of a thermoelectric generator set with heatsink for harvesting human body heat. *Energy Convers. Manag.* 2013, *68*, 260–265. [CrossRef]
5. Zaghari, B.; Weddell, A.S.; Esmaeili, K.; Bashir, I.; Harvey, T.J.; White, N.M.; Mirring, P.; Wang, L. High-temperature self-powered sensing system for a smart bearing in an aircraft jet engine. *IEEE Trans. Instrum. Meas.* 2020, *69*, 6165–6174. [CrossRef]
6. Pourkiaei, S.M.; Ahmadi, M.H.; Sadeghzadeh, M.; Moosavi, S.; Pourfayaz, F.; Chen, L.; Yazdi, M.A.P.; Kumar, R. Thermoelectric cooler and thermoelectric generator devices: A review of present and potential applications, modeling and materials. *Energy* 2019, *186*, 115849. [CrossRef]
7. Hu, S.; Chen, X. TEG—An Ideal Power Source of Gas Transmission Pipeline. *Oil Gas Storage Transport.* 1996, *12*, 22–26.
8. Glosch, H.; Ashauer, M.; Pfeiffer, U.; Lang, W. A thermoelectric converter for energy supply. *Sens. Actuators A Phys.* 1999, *74*, 246–250. [CrossRef]
9. Iannacci, J. Microsystem based energy harvesting (EH-MEMS): Powering pervasivity of the internet of things (IoT) e a review with focus on mechanical vibrations. *J. King Saud. Univ. Sci.* 2017, *31*, 66–74. [CrossRef]
10. Li, X.; Cai, K.; Gao, M.; Du, Y.; Shen, S. Recent advances in flexible thermoelectric films and devices. *Nano Energy* 2021, *89*, 106309. [CrossRef]
11. Tohidi, F.; Holagh, S.G.; Chitsaz, A. Thermoelectric Generators: A comprehensive review of characteristics and applications. *Appl. Therm. Eng.* 2022, *201*, 117793. [CrossRef]
12. Kim, T.Y.; Kwak, J.; Wook Kim, B. Energy harvesting performance of hexagonal shaped thermoelectric generator for passenger vehicle applications: An experimental approach. *Energy Convers. Manag.* 2018, *160*, 14–21. [CrossRef]
13. Shu, G.; Ma, X.; Tian, H.; Yang, H.; Chen, T.; Li, X. Configuration optimization of the segmented modules in an exhaust-based thermoelectric generator for engine waste heat recovery. *Energy* 2018, *160*, 612–624. [CrossRef]
14. Lan, S.; Yang, Z.; Chen, R.; Stobart, R. A dynamic model for thermoelectric generator applied to vehicle waste heat recovery. *Appl. Energy* 2018, *210*, 327–338. [CrossRef]
15. Li, B.; Huang, K.; Yan, Y.; Li, Y.; Twaha, S.; Zhu, J. Heat transfer enhancement of a modularised thermoelectric power generator for passenger vehicles. *Appl. Energy* 2017, *205*, 868–879. [CrossRef]
16. Agudelo, A.F.; García-Contreras, R.; Agudelo, J.R.; Armas, O. Potential for exhaust gas energy recovery in a diesel passenger car under European driving cycle. *Appl. Energy* 2016, *174*, 201–212. [CrossRef]
17. Ge, Y.; Liu, Z.; Sun, H.; Liu, W. Optimal design of a segmented thermoelectric generator based on three-dimensional numerical simulation and multiobjective genetic algorithm. *Energy* 2018, *147*, 1060–1069. [CrossRef]
18. Muralidhar, N.; Himabindu, M.; Ravikrishna, R.V. Modeling of a hybrid electric heavy duty vehicle to assess energy recovery using a thermoelectric generator. *Energy* 2018, *148*, 1046–1059. [CrossRef]
19. Zhang, D.; Qin, J.; Feng, Y.; Ren, F.; Bao, W. Performance evaluation of power generation system with fuel vapor turbine onboard hydrocarbon fueled scramjets. *Energy* 2014, *77*, 732–741. [CrossRef]
20. Cheng, K.; Qin, J.; Jiang, Y.; Lv, C.; Zhang, S.; Bao, W. Performance assessment of multi-stage thermoelectric generators on hypersonic vehicles at a large temperature difference. *Appl. Therm. Eng.* 2017, *130*, 1598–1609. [CrossRef]
21. Cheng, K.; Zhang, D.; Qin, J.; Zhang, S.; Bao, W. Performance evaluation and comparison of electricity generation systems based on single- and two-stage thermoelectric generator for hypersonic vehicles. *Acta Astronaut.* 2018, *151*, 15–21. [CrossRef]
22. Cheng, K.; Qin, J.; Jiang, Y.; Zhang, S.; Bao, W. Performance comparison of single and multi-stage onboard thermoelectric generators and stage number optimization at a large temperature difference. *Appl. Therm. Eng.* 2018, *141*, 456–466. [CrossRef]

Disclaimer/Publisher's Note: The statements, opinions and data contained in all publications are solely those of the individual author(s) and contributor(s) and not of MDPI and/or the editor(s). MDPI and/or the editor(s) disclaim responsibility for any injury to people or property resulting from any ideas, methods, instructions or products referred to in the content.

Article

Design of Thermoelectric Generators and Maximum Electrical Power Using Reduced Variables and Machine Learning Approaches

Alexander Vargas-Almeida [1], Miguel Angel Olivares-Robles [2,*] and Andres Alfonso Andrade-Vallejo [2]

[1] Departamento de Ingeniería en Automatización y Control Industrial, Universidad Politecnica del Golfo de Mexico, Carretera Federal Malpaso-El Bellote Km. 171, Paraíso 86800, Mexico; alexvargas.almeida@gmail.com
[2] Instituto Politecnico Nacional, SEPI ESIME-Culhuacan, Culhuacan, Ciudad de Mexico 04430, Mexico; ing.andres.andrade@gmail.com
* Correspondence: olivares@ipn.mx; Tel.: +52-57296000 (ext. 73262)

Citation: Vargas-Almeida, A.; Olivares-Robles, M.A.; Andrade-Vallejo, A.A. Design of Thermoelectric Generators and Maximum Electrical Power Using Reduced Variables and Machine Learning Approaches. *Energies* 2023, 16, 7263. https://doi.org/10.3390/en16217263

Academic Editors: Wei-Hsin Chen and Diana Enescu

Received: 4 May 2023
Revised: 26 August 2023
Accepted: 6 September 2023
Published: 26 October 2023

Copyright: © 2023 by the authors. Licensee MDPI, Basel, Switzerland. This article is an open access article distributed under the terms and conditions of the Creative Commons Attribution (CC BY) license (https://creativecommons.org/licenses/by/4.0/).

Abstract: This work aims to contribute to studies on the geometric optimization of thermoelectric generators ($TEGs$) through a combination of the reduced variables technique and supervised machine learning. The architecture of the thermoelectric generators studied, one conventional and the other segmented, was determined by calculating the cross-sectional area and length of the legs, and applying reduced variables approximation. With the help of a supervised machine learning algorithm, the values of the thermoelectric properties were predicted, as were those of the maximum electrical power for the other temperature values. This characteristic was an advantage that allowed us to obtain approximate results for the electrical power, adjusting the design of the $TEGs$ when experimental values were not known. The proposed method also made it possible to determine the optimal values of various parameters of the legs, which were the ratio of the cross-sectional areas (A_p/A_n), the length of the legs (l), and the space between the legs (H). Aspects such as temperature-dependent thermoelectric properties (Seebeck coefficient, electrical resistivity, and thermal conductivity) and the metallic bridge that connects the legs were considered in the calculations for the design of the $TEGs$, obtaining more realistic models. In the training phase, the algorithm received the parameter (H) and an operating temperature value as input data, to predict the corresponding value of the maximum power produced. This calculation was performed for conventional and segmented systems. Recent advances have opened up the possibility of applying an algorithm for designing conventional and segmented thermocouples based on the reduced variables approach and incorporating a supervised machine learning computational technique.

Keywords: thermoelectric generator design; dimensional parameters; maximum power; supervised machine learning

1. Introduction

The design of thermoelectric generators ($TEGs$) using geometric optimization techniques is the topic of interest in this work. Here, we describe some advantages of this type of design. First, it allows us to make better use of the thermoelectric material; to date, the best materials have been inorganic compounds (such as Bi_2Te_3), which have a relatively low abundance on Earth, and their manufacture requires a highly complex vacuum process. It is also possible to gradually adjust the dimensions of the system according to the space available for its coupling to a heat source. One parameter used in the geometric method is the cross-sectional area of the leg [1]. For example, it is possible to maximize the output power of the TEG as a function of the variable cross-section; in fact, in [2], it was stated that "The geometry of the TEG has a vital impact on the thermal resistance and the electrical resistance, influencing its integral performance".

Another aspect that motivated the development of this work is the emergence of new manufacturing techniques, such as additive manufacturing, which allow customizing thermoelectric systems according to the requirements of the field of application (from low-power applications, medical and wearable devices, Internet of Things, and wireless sensor networks; to high-power applications, industrial electronics, automotive engines, and aerospace). However, as rightly mentioned in [3], "little knowledge exists about which shapes are beneficial in applications with different thermal conditions". This author analyzed the effect of different thermoelectric leg designs on device performance. The authors in [4] proposed an optimized design of legs with special geometric shapes and their manufacture using 3D printing to increase the output power of *TEGs*.

In [2], two relevant aspects were mentioned: (a) in the current literature, there are still few studies on the optimization of the geometry of *TEGs*, especially for the shape of the legs; (b) recently, algorithms have been used to design optimal devices by simultaneously analyzing two or more geometric parameters.

The idea of geometric optimization comes from the structural aspect of the thermoelectric generator, which in its most elementary form (thermocouple) is composed of two legs of semiconductor material (for example, BiTe or SiGe), one *n*-type and the other *p*-type, which are electrically connected in series by means of a metallic bridge (for example, copper) (Figure 1); these legs have a rectangular prism shape. Then, two geometric parameters present in each leg are identified, the cross-sectional area (A) and the length (l). These geometric parameters are linked to two properties of thermoelectric materials, thermal conductance (K) and electrical resistance (R). This relationship between the geometric parameters and the thermoelectric properties is useful for adjusting the shape and size of the thermocouple. To achieve this dimensioning, a useful technique is the analysis of the electrical power produced by the TEG, in terms of the geometric parameters.

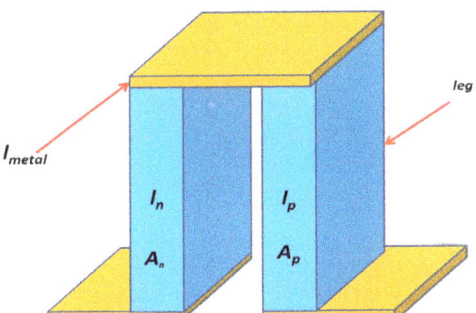

Figure 1. Design obtained for the conventional thermocouple.

The main motivation for carrying out this work arose from identifying that currently, in the thermoelectricity field, there is a growth in the amount of researchers interested in developing *TEGs* devices for harvesting energy from waste heat sources [5,6]. To achieve the maximum use of this heat, it is essential to analyze the geometric characteristics of the device. As previously mentioned, geometric optimization methods require knowledge of thermoelectric properties, which are linked to the dimensional parameters and that also vary with temperature. These conditions require knowledge of property measurements to achieve the best adjusted *TEG* design for the requirements imposed by the heat source, the available space, and the load resistance of the system that will use the power produced by the *TEGs*. However, not all researchers attempting to design *TEGs* have a materials laboratory or the equipment required to prepare samples and measurements. At present, a useful resource is simulation software to study the properties of materials and design *TEGs*; even so, depending on the type of software, investment in licensing is required, in addition to high-end computer equipment.

Seeking to develop an affordable alternative for the community intending to design $TEGs$, in this work, we propose developing a methodology built on three principles:

(I) Use data from experimental measurements found in publications by specialists in thermoelectric materials; (II) Use a formalism or an approximation that allows the immediate use of the data obtained from the literature to obtain the leg dimensions (An, Ap, l); (III) Merge the selected formalism or approximation with a prediction algorithm that allows us to generate the design best adjusted to the operating conditions of the environment.

The three previous principles guided us toward combining the reduced variables technique with supervised machine learning (SML) in a design process that consists of the following steps: (a) take advantage of the qualities of the reduced current approximation to obtain the architecture of the thermoelectric generator by calculating the parameters (cross-sectional area and length of its legs); (b) with the help of an algorithm (SML), the values of the thermoelectric properties and the maximum electrical power (P) are predicted for any temperature value; and (c) the values obtained in the prediction are useful for adjusting the design for the operating conditions. In addition, the algorithm has the outstanding feature of simultaneously analyzing and determining various parameters related to the geometry of the legs that maximize (P). These parameters are the cross-sectional area ratio (Ap/An), length (l), length n ratio (ln_2/ln_1), length p ratio (lp_2/lp_1) (in the case of a segmented TEG), and the space between legs (H).

The results provide useful information for the construction of optimal devices and their possible applications. The scope of this work was extended to a new training phase, in which it is possible to introduce values of the parameter H and of the temperature, managing to predict the corresponding value of the maximum power. Of course, this study has limitations, which are discussed in later sections; however, these first results allowed us to determine that it is possible to obtain an algorithm for the design of conventional and segmented thermocouples based on a reduced variables approach fused with a supervised machine learning calculation model, trained for various thermoelectric materials. The utility and scope of the method are shown when confronted with a computerized and experimental model in Section 6.

Notes on the State of the Art

The reduced variables approach is a strategy that can be applied for the optimization of power, efficiency, and even the coefficient of performance, modifying the current flow in the legs by adjusting the cross-sectional area. This feature allows us to model the architecture of the TEG, managing the size of the cross-sectional area and length of the legs. The scope of this tool was reported in the works of [7–10].

On the other hand, in the field of thermoelectricity, machine learning has been applied for the prediction of material properties and for the design of new materials. As [11] rightly mentioned, "the machine learning technique can provide a powerful discovery tool for thermoelectric materials with respect to the new chemical composition, nanostructural design, stoichiometry optimization, etc.". Specifically, supervised machine learning, which is based on algorithms that learn from an input dataset and a training dataset and manage to predict unseen data or future values—divided into two categories of classification and regression [12]—has been applied to carry out the synthesis of a new material spin-driven thermoelectric effect (STE) [13].

2. Related Work

In the search for a sustainable society and world, thermoelectric generators are an alternative that can contribute to the economic and social development of regions where it is possible to take advantage of the products generated by biodiversity; for example, in [14], the coupling of a thermoelectric generator in a lignin biorefinery was reported (which is a heteropolymer that is part of the cell wall of the vascular tissue of plants, one of the main components of the biomass that can be used to obtain renewable products as raw material for biofuels). The use of thermoelectric generators for the recovery of waste heat from biomass stoves was reported in [15]; in this work, the voltage produced by a thermoelectric generator placed between the hot wall of a stove and a heatsink was investigated. The voltage produced by this system was measured and compared for different biomass fuels, such as wood chips, walnut shells, cobs of corn, and coconut shells. Their results showed that each biomass fuel had its own combustion characteristics that allowed the TEG to generate a high voltage at specific values. It is also very interesting to comment on the work of [16], in which a self-powered marine mammal condition monitoring system was proposed based on the hybrid energy-harvesting mode of a triboelectric nanogenerator ($TENG$) and a thermoelectric microgenerator ($MTEG$). Table 1 below shows some very interesting works on the design of $TEGs$ that are related to the one presented in this article.

Table 1. Related work.

Reference	Characteristics
[1]	A thermal-electric coupled mixed-method algorithm that predicts TEG performance and optimizes the crosssectional area along the leg length in order to optimize power output or thermal conversion efficiency
[2]	This work proposes an optimization study to maximize the output power of variable cross-section $TEGs$ for solar energy utilization by coupling finite element method (FEM) and optimization algorithm. Six geometric variables along with the external load resistance are optimized by genetic algorithm (GA) and particle swarm optimization (PSO).
[3]	Various leg shapes (rectangular prisms, prisms with interior hollows, trapezoids, hourglass, and Y-shape) were modeled numerically to determine their thermal and electrical performance under constant temperature and heat flux boundary conditions. Two thermoelectric materials, bismuth telluride and silicon germanium, were modeled to capture both low and high temperature application cases, respectively.
[4]	This work proposes a new geometric design concept to improve the output voltage and power of the TE legs in RTGs based on increasing side area to enhance heat dissipation caused by convective heat transfer and radiative heat transfer. Helix-shaped and spoke-shaped TE legs with different geometrical shapes are designed.

3. Materials and Methods

3.1. Conventional Thermocouple

Dimensional Parameters

The parameters used for the design of thermoelectric legs were the cross-sectional area (A_n), (A_p) and the length $(l_p = l_n = l)$. Snyder [10] rigorously formulated the reduced variable approximation. A critical parameter of this approximation is the reduced (relative) current, defined as:

$$u = \frac{J}{\kappa \nabla T} \quad (1)$$

where κ is the thermal conductivity and T is the absolute temperature.

Snyder [10] proposed an iteration equation to calculate the numerical values of the quantity $(u\kappa)$ required for calculations. The $(u\kappa)$ values were obtained from tables consulted in [10] for different temperature values. This quantity is the kernel of the integrals that appear in Equations (4) and (12), which were used to calculate the cross-sectional area and length of the legs. Below is the detailed procedure for calculating (A_n), (A_p), and (l), applying reduced variable approximation, which are the main parameters for TEG design.

The first step in the reduced variable procedure is to establish the equal lengths of both legs; that is, $(l_p = l_n)$:

$$I = J_p A_p = J_n A_n \quad (2)$$

if the current density (Jl) is defined as follows:

$$Jl = \int_{T_c}^{T_h} \kappa u \, dT \tag{3}$$

where κ is the thermal conductance and u is the reduced current density. For a complete understanding of the equations of the reduced variable formalism used in this article, we suggest consulting reference [10]. Combining Equations (2) and (3), the following relationship is obtained:

$$\frac{A_p}{A_n} = \frac{-J_n}{J_p} = \frac{-\int_{T_c}^{T_h} u_n \kappa_n \, dT}{\int_{T_c}^{T_h} u_p \kappa_p \, dT} \tag{4}$$

To calculate the integrals of Equation (4), data from Table 2 are required; these were obtained from the literature; see reference [10].

Table 2. Numerical data ($u\kappa$), material (Bi_2Te_3).

T (K)	$u_p k_p dT$ (A/cm)	$u_n k_n dT$ (A/cm)
$T_0 = 298$	$u_p k_p(T_0) = 0.8132$	$u_n k_n(T_0) = -0.4966$
$T_1 = 323$	$u_p k_p(T_1) = 0.8350$	$u_n k_n(T_1) = -0.5206$
$T_2 = 348$	$u_p k_p(T_2) = 0.8424$	$u_n k_n(T_2) = -0.5679$
$T_3 = 373$	$u_p k_p(T_3) = 0.8435$	$u_n k_n(T_3) = -0.6312$
$T_4 = 398$	$u_p k_p(T_4) = 0.8466$	$u_n k_n(T_4) = -0.6933$
$T_5 = 423$	$u_p k_p(T_5) = 0.8603$	$u_n k_n(T_5) = -0.4454$

The integrals $-\int_{T_c}^{T_h} u_n \kappa_n \, dT$ and $\int_{T_c}^{T_h} u_p \kappa_p \, dT$ are calculated applying the Newton–Cotes method (fourth order). The results for each integral are shown:

$$-\int_{T_c}^{T_h} u_n \kappa_n \, dT = -57.78 \tag{5}$$

$$\int_{T_c}^{T_h} u_p \kappa_p \, dT = 83.82 \tag{6}$$

The area ratio is

$$\frac{A_p}{A_n} = 0.69 \tag{7}$$

Now, it is possible to calculate the current densities in the legs;

$$J_p = \frac{U_{total-h}}{A_{total}} \frac{1 + \frac{A_n}{A_p}}{\Phi_{p-h} - \Phi_{n-h}} \tag{8}$$

Φ_{p-h} and Φ_{n-h} are the thermoelectric potentials at temperature T_h = 423 K;

$$\frac{U_{total-h}}{A_{total}} = 20 \, (\text{W}/\text{cm}^2)$$
$$\Phi_{ph} = 0.37 \, (\text{V})$$
$$\Phi_{nh} = -0.49 \, (\text{V}) \tag{9}$$

At these values, the current density in the p-type leg is

$$J_p = 57.32 \, \text{mA}/\text{cm}^2 \tag{10}$$

For the current density in the n-type leg, the value J_p and quotient (7) are used;

$$J_n = 39.51 \text{ mA/cm}^2 \tag{11}$$

Now, it is possible to know the length (l), for which the following equation is applied:

$$\int_{T_c}^{T_h} k_{n,p} u_{n,p} dT = J_{n,p} l_{n,p} \tag{12}$$

The next value of l is obtained;

$$l = 1.46 \text{ (mm)} \tag{13}$$

To calculate the cross-sectional areas A_p and A_n, A_{total} is defined as the total area. For calculation purposes, $A_{total} = 1 \text{ mm}^2$ is selected. The results are

$$A_p = 0.41 \text{ mm}^2$$
$$A_n = 0.59 \text{ mm}^2 \tag{14}$$

Now, the thickness (l_{metal}) of the metal bridge is added. Figure 1 shows a sketch of the design obtained:

This first system can be used as a basic unit for the design of a thermoelectric module composed of several thermocouples.

4. Temperature-Dependent Thermoelectric Properties and Supervised Machine Learning

Integrals [5,6] could be calculated thanks to the fact that we know the measurements of the thermoelectric properties at the temperatures in Table 2, generating a first TEG design with their respective parameters (A_n, A_p, l). However, the system obtained is specific for that temperature range, in such a way that it is not possible to adjust the design for a wider or narrower range. Trying to make a new design may require preparing a new material sample for new measurements or perhaps using advanced software for numerical design and simulation; there is a possibility that researchers do not have some of these resources. This difficulty can be overcome with the implementation of a supervised machine learning algorithm that allows predicting the values of the Seebeck coefficient, α; electrical resistivity, ρ; and thermal conductivity, κ, properties at any temperature value. Table 3 shows the measurements of these properties for the $BiTe$ material.

The data in Table 3 were used to train the prediction code, for which an 80/20 rule was used; that is, 80% of the data were used for the training set and 20% for the test set. The code was generated using the Wolfram Mathematica 13.2 software using a predictive function that has the advantage of automatically selecting the most appropriate prediction model according to the behavior of the experimental data used for training. Specifically, the algorithm was applied to obtain a set of 50 data points for each thermoelectric property; however, more data can be predicted for any established temperature range. The distance between the data was reduced to 2.5 K. Figure 2 shows a spreadsheet containing the prediction results of the thermoelectric properties generated by the algorithm for each of the materials selected for the design of the conventional generator.

Table 3. Experimental data of the thermoelectric properties of the Bi_2Te_3 material.

T (K)	α_p (µV/K)	ρ_p (10^{-3} Ωcm)	κ_p (mW/cm K)
$T_0 = 298$	173	0.927	9.63
$T_1 = 323$	185	1.015	9.85
$T_2 = 348$	194	1.198	9.87
$T_3 = 373$	200	1.415	9.79
$T_4 = 398$	203	1.632	9.70
$T_5 = 423$	204	1.834	9.71
	α_n (µV/K)	ρ_n (10^{-3} Ωcm)	κ_n (mW/cm K)
	−209	2.38	8
	−213	2.61	8.23
	−210	2.79	8.72
	−201	2.90	9.8
	−187	2.94	10.92
	−171	2.92	12.07

Temperature (K)	Properties BiTe (p)			Properties BiTe (n)		
	Seebeck coefficient	electrical resistivity	Thermal conductivity	Seebeck coefficient	electrical resistivity	Thermal conductivity
298	176.441	0.907216	9.7503	−210.156	2.4578	7.7946
300.5	177.061	0.925053	9.78009	−209.577	2.4687	7.87509
303	177.681	0.942891	9.82479	−208.999	2.47959	7.95558
305.5	178.302	0.960728	9.88438	−208.421	2.49049	8.03606
308	178.922	0.978566	9.89928	−207.842	2.50139	8.11655
310.5	179.542	0.996404	9.94397	−207.264	2.51229	8.19704
313	180.163	1.01424	9.95887	−206.685	2.52319	8.27752
315.5	180.783	1.03208	9.97377	−206.107	2.53409	8.35801
323	182.644	1.08559	10.0185	−204.372	2.56678	8.59947
325.5	183.264	1.10343	10.0334	−203.793	2.57768	8.67996
328	183.885	1.12127	10.0483	−203.215	2.58858	8.76044
330.5	184.505	1.1391	10.0632	−202.637	2.59948	8.84093
333	185.125	1.15694	10.0781	−202.058	2.61037	8.92142
335.5	185.746	1.17478	10.093	−201.48	2.62127	9.0019
338	186.366	1.19262	10.1079	−200.901	2.63217	9.08239
340.5	186.986	1.21046	10.1228	−200.323	2.64307	9.16288
343	187.607	1.22829	10.1377	−199.745	2.65397	9.24336
345.5	188.227	1.24613	10.1526	−199.166	2.65397	9.32385
348	188.847	1.26397	10.1674	−198.588	2.67576	9.40434
350.5	189.468	1.28181	10.1823	−198.01	2.68666	9.48482
353	190.088	1.29964	10.1972	−197.431	2.69756	9.56531

Figure 2. Spreadsheet containing the 50 data generated by the prediction algorithm.

This algorithm helps predict a material's thermoelectric property value for any operating temperature value. As shown below, it is possible to calculate the maximum electrical power of a conventional system for any T value within this range.

4.1. Segmented Thermocouple

As a result of efforts to take advantage of heat in a wide temperature range, the segmentation technique was developed. It consists of joining segments of various thermoelectric materials and allowing thermal and electrical contact between them; its principle is based on the fact that each of the segments will be subjected to the temperature range in which it reaches its highest figure of merit value.

4.1.1. Design of a Segmented Thermocouple

The materials selected were Bi_2Te_3 and Zn_4Sb_3 (p-type) and Bi_2Te_3 and $CoSb_3$ (n-type). The operating temperatures selected were $T_C = 398$ K and $T_H = 573$ K. Tables 4 and 5 show the values of the product $u k$ of the materials Bi_2Te_3 and Zn_4Sb_3 (p-type), respectively.

Table 4. Numerical data of the $u\kappa$ product of the material Bi_2Te_3.

T (K)	$u\kappa_{Bi_2Te_3}$ (1/V)
398	0.8466
423	0.8603
448	0.8932
473	0.3339

Table 5. Numerical data of the $u\kappa$ product of the material Zn_4Sb_3.

T (K)	$u\kappa_{Zn_4Sb_3}$ (1/V)
482	0.3167
498	0.4999
523	0.5107
548	0.5257
573	0.5430

Tables 6 and 7 show the values of the product uk of the materials Bi_2Te_3 and $CoSb_3$ (n-type), respectively.

Table 6. Numerical data of the $u\kappa$ product of the material Bi_2Te_3.

T (K)	$u\kappa_{Bi_2Te_3}$ (1/V)
398	−0.6933
423	−0.4454

Table 7. Numerical data of the $u\kappa$ product of the material $CoSb_3$.

T (K)	$u\kappa_{CoSb_3}$ (1/V)
440	−0.7948
448	−2.4649
473	−2.4387
498	−2.4164
523	−2.3984
548	−2.3855
548	−2.3783
573	−2.3783

Applying the treatment defined in the previous sections, the following parameters of the segmented thermocouple were obtained:

Table 8 contains the design parameters of the segmented thermoelectric generator. A design sketch is shown in Figure 3.

Table 8. Current density and dimensional parameters of the segments.

Parameter	Bi_2Te_3 (p)	Zn_4Sb_3 (p)	Bi_2Te_3 (n)	$CoSb_3$ (n)
J (mA/cm^2)	29.16	29.16	96.96	96.96
l (mm)	1.76	1.56	0.10	3.21
A (mm^2)	0.77	0.77	0.23	0.23

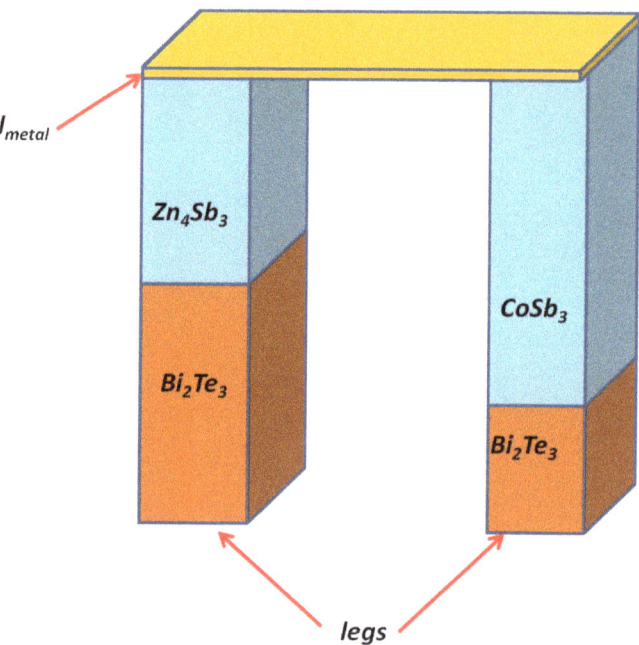

Figure 3. Design obtained for the segmented thermocouple.

Figure 4 shows a diagram that represents the interaction between the components in the system.

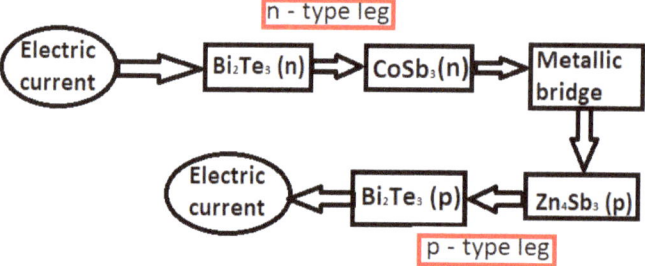

Figure 4. Sequence diagram showing the interaction between segmented thermocouple components.

Again, the prediction algorithm was applied to obtain 50 data points for each of the thermoelectric properties. For each of the materials selected in the design, the distance between the data points was reduced to 2.5 K. Figure 5 shows a spreadsheet containing the prediction results of the thermoelectric properties generated by the algorithm for each of the materials selected for segmented generator design.

This algorithm helps predict the values of the materials' thermoelectric properties for any operating temperature value within the range of 398–573 K. As shown below, it is possible to calculate the maximum electrical power of the segmented system for any T value within this range.

Temperature (K)	properties BiTe (p)			properties BiTe (n)			propertiess ZnSb			properties CoSb		
	Seebeck coefficient	electrical resistivity	Thermal conductivity	Seebeck coefficient	electrical resistivity	Thermal conductivity	Seebeck coefficient	electrical resistivity	Thermal conductivity	Seebeck coefficient	electrical resistivity	Thermal conductivity
398	203.446	1.64027	9.76891	-210.423	2.82185	11.0343	149.572	2.42363	5.93	-144.234	0.756025	43.2601
401.5	203.191	1.66514	9.79564	-208.509	2.83132	11.1526	150.221	2.43296	5.93	-145.032	0.759071	43.1488
405	202.935	1.69001	9.82237	-206.596	2.84078	11.2709	150.871	2.44228	5.93	-145.83	0.762117	43.0376
408.5	202.679	1.71488	9.84909	-204.682	2.85024	11.3892	151.521	2.45161	5.93	-146.628	0.765163	42.9264
412	202.424	1.73975	9.87582	-202.768	2.8597	11.5075	152.17	2.46093	5.93	-147.425	0.768209	42.8152
415.5	202.168	1.76463	9.90255	-200.854	2.86917	11.6258	152.82	2.47026	5.93	-148.223	0.771255	42.704
419	201.913	1.7895	9.92927	-198.941	2.87863	11.744	153.47	2.47958	5.93	-149.021	0.774301	42.5928
422.5	201.657	1.81437	9.956	-197.027	2.88809	11.8623	154.12	2.48891	5.93	-149.819	0.777346	42.4816
426	201.402	1.83924	9.98273	-195.113	2.89756	11.9806	154.769	2.49823	5.93	-150.617	0.780392	42.3703
429.5	201.146	1.86411	10.0095	-193.199	2.90702	12.0989	155.419	2.50756	5.93	-151.415	0.783438	42.2591
433	200.89	1.88898	10.0362	-191.286	2.91648	12.2172	156.069	2.51688	5.93	-152.213	0.786484	42.1479
436.5	200.635	1.91385	10.0629	-189.372	2.92595	12.3354	156.719	2.52621	5.93	-153.011	0.78953	42.0367
440	200.124	1.93872	10.0896	-187.458	2.93541	12.4537	157.368	2.53553	5.93	-153.809	0.792576	41.9255
443.5	199.868	1.96359	10.1164	-185.544	2.94487	12.572	158.018	2.54486	5.93	-154.607	0.795622	41.8143
447	199.613	1.98846	10.1431	-183.631	2.95433	12.6903	158.668	2.55418	5.93	-155.405	0.798668	41.7031
450.5	199.357	2.01333	10.1698	-181.717	2.9638	12.8086	159.318	2.56351	5.93	-156.202	0.801714	41.5919
454	199.101	2.0382	10.1965	-179.803	2.97326	12.9268	159.967	2.57283	5.93	-157	0.80476	41.4806
457.5	198.846	2.06307	10.2233	-177.889	2.98272	13.0451	160.617	2.58216	5.93	-157.798	0.807806	41.3694
461	198.59	2.08794	10.25	-175.976	2.99219	13.1634	161.267	2.59148	5.93	-158.596	0.810852	41.2582

Figure 5. Spreadsheet containing the 50 data points generated by the prediction algorithm for the segmented thermocouple.

4.1.2. Evaluation of the Feasibility of the Segmented Thermocouple by Means of the Compatibility Factor

An important aspect to remember in the design of segmented thermocouples is that only certain combinations of materials are appropriate, because there is a risk of building a thermocouple with a low efficiency, even lower than that of a conventional thermocouple. A useful resource for evaluating combinations of materials is the compatibility factor (S) [10], defined as

$$S = \frac{\sqrt{1 + Z\overline{T}} - 1}{\alpha \overline{T}} \quad (15)$$

Applying Equation (15), it can be confirmed that the materials selected in the system studied in this work were correct for the segmentation; see Table 9. In this table, it can be observed that the values of S between p-type and n-type materials differed by a factor not greater than 2, as indicated by the rule.

Table 9. Values of the compatibility factor S for each of the materials selected for segmentation.

Material	S
$Bi_2Te_3\ (p)$	4.22
Zn_4Sb_3	4.383
$Bi_2Te_3\ (n)$	2.27
$CoSb_3$	2.26

5. Results and Discussion

5.1. Maximum Electrical Power of a Conventional Thermocouple

As is well known in the field of thermoelectricity, the maximum power (P_{max}) of the generator is reached with the condition $R_{load} = R_{internal}$. In this work, an analysis of the maximum power of generators, conventional and segmented, was conducted for various physical conditions established for the dimensional parameters (cross-sectional area, length of the legs, space between the legs). First, the maximum power analysis is shown for the conventional system and later for the segmented system.

The maximum power of a conventional TEG is

$$P_{max-conventional} = \frac{(\alpha_p - \alpha_n)^2 (T_H - T_C)^2}{4(R_n + R_p + R_{met})}. \quad (16)$$

$P_{max-conventional}$ was analyzed for three conditions, as a function of (a) the space between the legs (H), (b) the ratio of cross-sectional areas (A_p/A_n), and (c) the length of the legs (l). Equation (16) was written for each of these parameters using relationships $R_n = \frac{\rho_n l}{A_n}$, $R_p = \frac{\rho_p l}{A_p}$, and $R_{metal} = \frac{\rho_{metal} l}{A_{metal}}$.

The electrical resistance (R) metal used in this study is presented in Table 10. Notice that for calculation purposes, a thickness of the metal bridge of 0.01 cm was selected.

Table 10. Electrical resistance metal bridge.

Component	Electrical Resistance (R)
Metal	$\frac{0.00002(0.01)}{A_{metal}}$

We have chosen not to present the equations, since our main objective was to analyze the behavior of the power curves. The results obtained for each one are shown below.

5.1.1. Maximum Power of the Conventional Generator: Space between the Legs

The space between the thermoelectric legs is considered a variable parameter; see Figure 6.

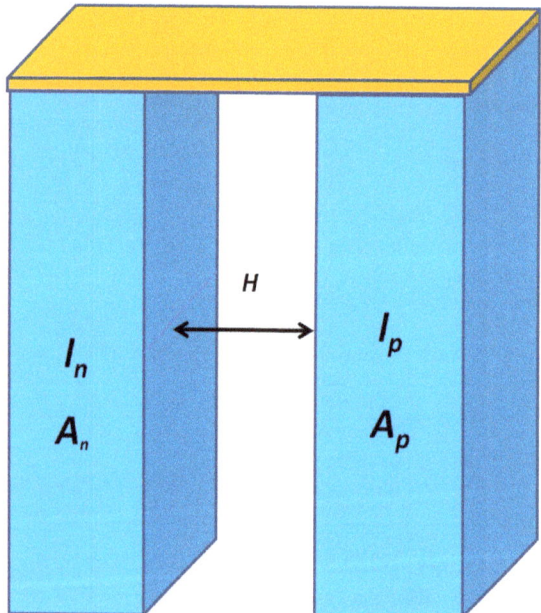

Figure 6. H: space between the thermoelectric legs.

Figure 7 shows the maximum electrical power as a function of the space between the legs for different values of temperature using the prediction algorithm. The values 298 K, 318 K, 348 K, 398 K, and 418 K were selected.

Figure 7 shows that the maximum power increased significantly in the range from 0 to 1 cm, after which the variation was insignificant. This is because the closer the legs are together, the greater the temperature difference between the hot and cold sides of the generator. It was also observed that the power values increased as the temperature decreased.

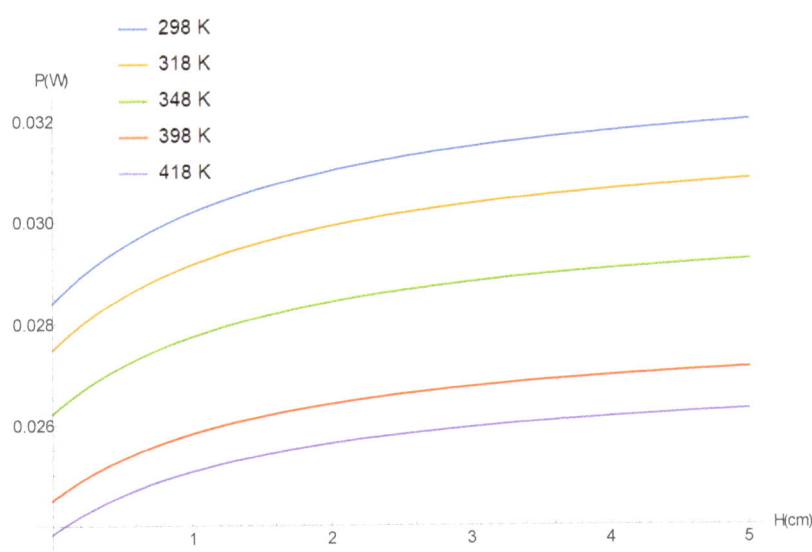

Figure 7. Maximum electrical power of the conventional thermoelectric generator as a function of the space between the legs, H.

5.1.2. Maximum Power of the Conventional Generator: Area Ratio

It is also helpful to observe how the maximum power varies concerning the ratio of areas, to determine the optimum value of the A_p and A_n areas, which must be maintained for the system to achieve the highest possible power value.

Figure 8 shows the region of rapid increase in each electrical power curve. The electrical power quickly grew until reaching close to 0.5; after that point, the variation was insignificant. Again, the maximum power was more significant for temperatures close to 298 K.

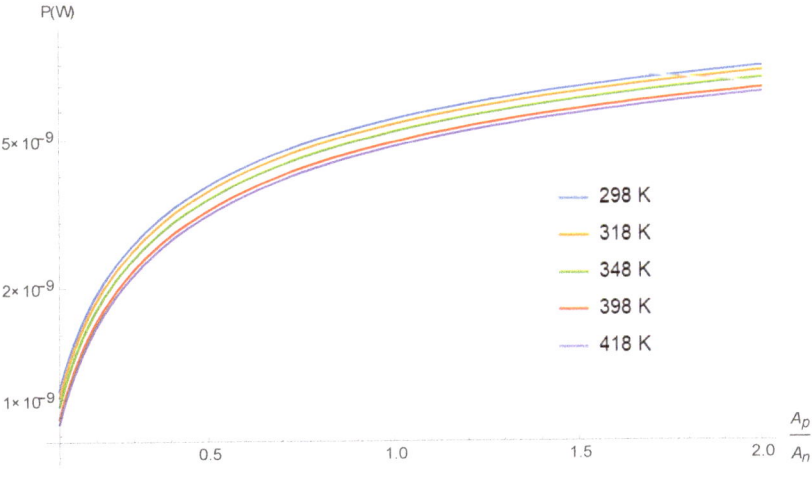

Figure 8. Maximum electrical power of the conventional thermoelectric generator as a function of area ratio (A_p/A_n).

5.1.3. Maximum Power of the Conventional Generator: Length of the Legs

From the variation in the maximum power as a function of the length of the legs, it was possible to determine the optimum length for the highest performance of the thermoelectric generator.

Figure 9 shows that for the selected temperatures, length values less than 0.1 cm would be appropriate to achieve the maximum performance.

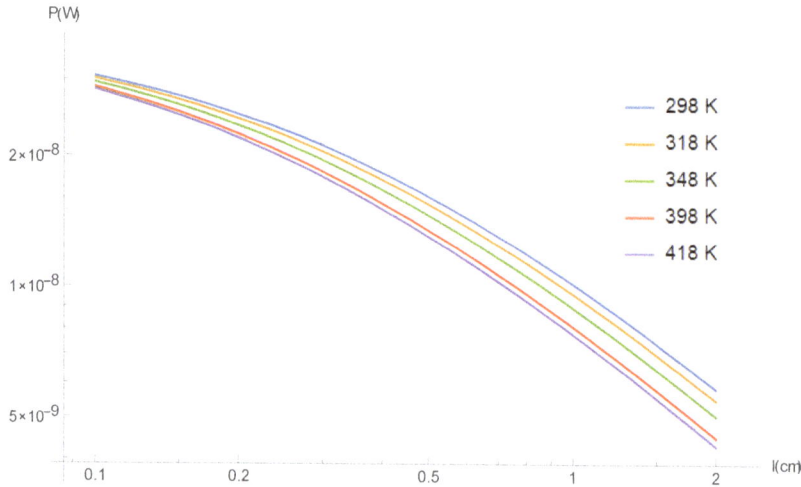

Figure 9. Maximum electrical power of the conventional thermoelectric generator as a function of the length of the legs.

5.2. Maximum Electrical Power of a Segmented Thermocouple

The maximum power produced by a segmented thermoelectric generator is defined using the following equation:

$$P_{max-segmented} = \frac{(\alpha_{p-effective} - \alpha_{n-effective})^2 (T_H - T_C)^2}{4(R_{n1} + R_{n2} + R_{p1} + R_{p2} + R_{met})} \quad (17)$$

Note that the effective Seebeck coefficient can be defined for each of the n-type and p-type segmented legs as

$$\alpha_{p-effective} = \frac{\kappa_{p1}\alpha_{p2} + \kappa_{p2}\alpha_{p1}}{\kappa_{p1} + \kappa_{p2}} \quad (18)$$

$$\alpha_{n-effective} = \frac{\kappa_{n1}\alpha_{n2} + \kappa_{n2}\alpha_{n1}}{\kappa_{n1} + \kappa_{p2}} \quad (19)$$

The electrical power produced by the segmented thermoelectric generator was analyzed for four conditions, which were (a) as a function of the space between the legs, (H), (b) as a function of the ratio of cross-sectional areas, (A_p/A_n), (c) as a function of the ratio of the lengths of p-type materials (l_{p2}/l_{p1}), and (d) as a function of the ratio of lengths of n-type materials (l_{n2}/l_{n1}). Similarly to the case of the conventional thermocouple, Equation (17) was written for each of the four parameters. Again, the key was the electrical resistance values, which in this case were (R_{n1}), (R_{n2}), (R_{p1}), (R_{p2}), and (R_{met}). Thus, the power curves for each were generated.

5.2.1. Maximum Power of the Segmented Thermoelectric Generator: Space between the Legs

Figure 10 shows that the system's maximum power was practically constant as the space between the legs changed, but this depended dramatically on the selected temperatures. It can also be seen that the temperature that produced the highest power was 510 K. This is interesting news for researchers and engineers who are working on developing new ways to generate power. The results of this study suggest that by carefully controlling the temperature, we could produce a constant and reliable source of power, regardless of the distance between the legs of the system. This could have a major impact on the development of new technologies, such as portable electronics and renewable energy sources.

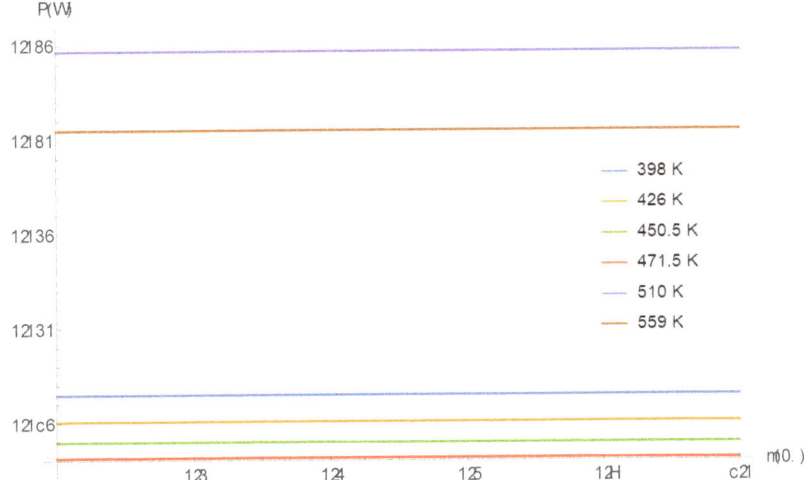

Figure 10. Maximum electrical power of the segmented thermoelectric generator as a function of the space between the legs (H).

5.2.2. Maximum Power of the Segmented Thermoelectric Generator: Area Ratio (A_p/A_n)

Figure 11 shows the maximum power for the segmented thermoelectric generator as a function of the area ratio. An attractive characteristic is the different intersections of the curves in the interval from 0 to 0.5. These intersection points are essential because they represent the same maximum power at different temperatures and at the same area ratio. This behavior results from the segmentation because, as it is well known, certain thermoelectric materials are more efficient than others at different temperatures. As the values of the area ratio after that interval increased, all the curves remained without notable changes. This analysis shows that the maximum power output of a segmented TEG can be increased by increasing the area ratio and operating the TEG at a temperature of 510 K.

5.2.3. Maximum Power of the Segmented Thermoelectric Generator: n-Type Length Ratio

In the case of the conventional thermocouple, analyzing the power variation as a function of the leg lengths is a simple task because $lp = ln$. However, for the case of a segmented TEG, the lengths are different for each segment; that is, a set of four lengths is formed (l_{n1}, l_{n2}, l_{p1}, and l_{p2}). A practical method is to analyze the electrical power as a function of the length ratios (l_{n2}/l_{n1}) and (l_{p2}/l_{p1}). Figure 12 shows the power of the segmented system as a function of the ratio (l_{n2}/l_{n1}) at different temperatures. The power was maximum for small values of (l_{n2}/l_{n1}) or when ($l_{n2} < l_{n1}$), but it decreased when this ratio increased; that is, ($l_{n2} > l_{n1}$). The temperature that showed the highest power value was $T = 426$ K. This result suggests that the segmented system was more efficient at generating power at small values of n-type length ratio. Notice that the electrical power decreased rapidly from the value 0.5.

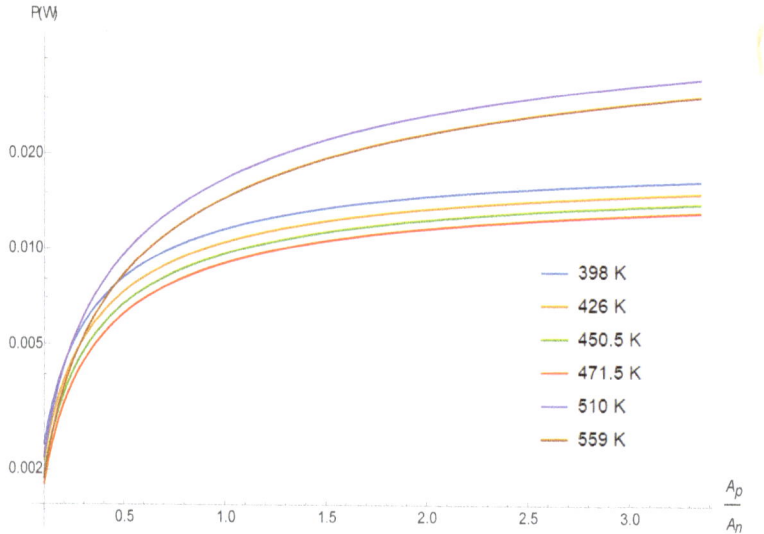

Figure 11. Maximum electrical power of the segmented thermoelectric generator as a function of the area ratio (A_p/A_n).

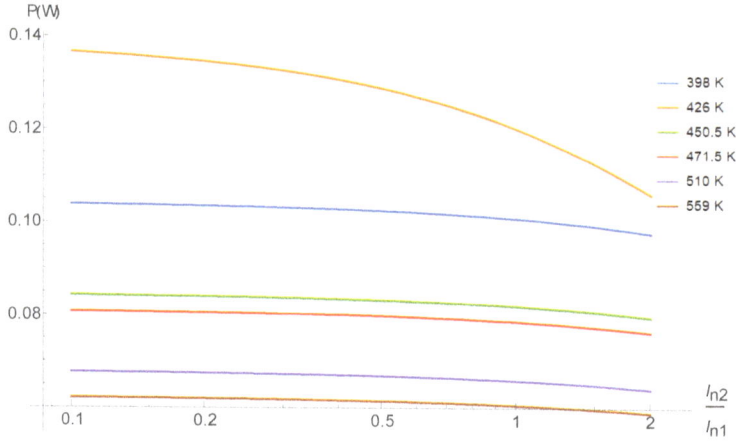

Figure 12. Maximum electrical power of the segmented thermoelectric generator as a function of the n-type leg ratio (l_{n2}/l_{n1}).

5.2.4. Maximum Power of the Segmented Thermoelectric Generator: p-Type Length Ratio

Figure 13 shows the maximum power as a function of the ratio (l_{p2}/l_{p1}) for different temperatures. An important feature is that the electrical power decreased rapidly for all intervals of values of the p-type leg ratio (l_{p2}/l_{p1}). This is because longer legs have more resistance, which limits the current flow and therefore the power output. Notice that the maximum power values of the system in this case were lower than those achieved with the quotient (l_{n2}/l_{n1}). These results suggest that for maximum electrical power, semiconductors should be segmented in n-type legs. This can reduce the resistance of the legs and allow for more current to flow, resulting in a higher power output. The highest maximum power output was achieved at temperatures around 398 K. The electrons in the semiconductor were more mobile at higher temperatures.

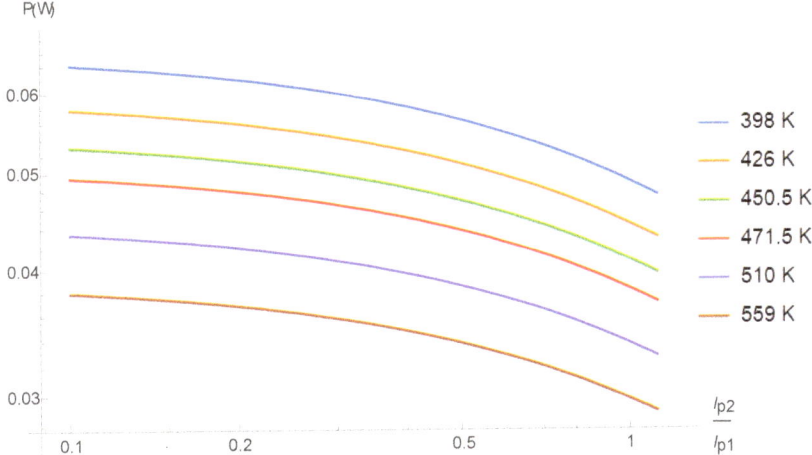

Figure 13. Maximum electrical power of the segmented thermoelectric generator as a function of the *p*-type leg ratio (l_{p2}/l_{p1}).

When comparing the two systems, conventional and segmented, regarding the space parameter (H) between the legs, the changes in maximum power values are more noticeable in the curves of the conventional thermocouple. In contrast, for the segmented thermocouple, the variations are minimal. The maximum power stabilized after reaching a maximum value at some specific value of H, because this parameter is related to the area of the metal bridge; thus, when H increased, the resistance of the metal bridge decreased and approaches zero. In this way, the maximum power value approached a constant value. In the case of the segmented thermocouple, it can be seen from Figure 10 that a smaller value of H ($0 \leq H \leq 1$) was required to reach the maximum power and stability, compared to the value of H required for the conventional thermocouple ($1 \leq H \leq 2$); see Figure 6. When analyzing the variation in power as a function of (A_p/A_n), both systems reached maximums at similar points ($A_p/A_n \approx 0.5$), after which it was no longer necessary to continue increasing the ratio. It is very interesting to observe the intersections between the power curves in Figure 11; it is possible that this was a consequence of the segmentation technique used (remember that each segment of thermoelectric material reached its maximum performance value for a specific temperature). Moreover, the intersection point of two or more of these curves indicates that at those temperatures the *TEG* was operating with the segments at maximum efficiency. For example, in the interval $0 < A_p/A_n \leq 0.5$, the power curve at 559 K intersects with the curves at 398 K, 426 K, and 450.5 K. When comparing the points of intersection with the temperature values of Tables 4–7, the following correspondence is found: "for a ratio of ($Ap/An = 0.5$), the optimal temperature values of each material are shown in Table 11".

Table 11. Optimum temperature values for each segment.

T (K)	Material
559	Zn_4Sb_3 (p)
450.5	$CoSb_3$ (n)
426	Bi_2Te_3 (p)
398	Bi_2Te_3 (n)

For the analysis of the maximum power as a function of the leg length (conventional TEG) or the ratio of lengths (segmented TEG), the Figures 9, 12 and 13 suggest that small values of these parameters favor the performance of the systems, because it is possible to reduce the electrical resistance.

The analysis of the temperatures in each case mentioned above showed that for the conventional thermocouple, the dominant temperature was 298 K. This result confirmed that the BiTe material reached its maximum figure of merit value at an operating temperature between 200 and 300 K. On the other hand, regarding the segmented thermoelectric generator, Figures 10 and 11 show that the temperature that produced the highest power was 510 K. However, in Figure 12, the segmented system produced the highest maximum power value at 426 K, while Figure 13 shows that at $T = 398$ K, the system generated a high maximum power value. These peculiarities show that, unlike the conventional thermocouple, in the case of the segmented thermocouple, there was a greater sensitivity to changes in the operating temperature, in a specific way concerning the condition established for certain dimensional parameters.

6. Model Building and Experimental Setup

In order to verify the validity of the proposed methodology, in this section, we consider the results of the work of Crane et al. [17], who created a design using a computerized model and performed an experiment with the built prototype of a system called a three-couple TEG engine; see Figure 14 [17].

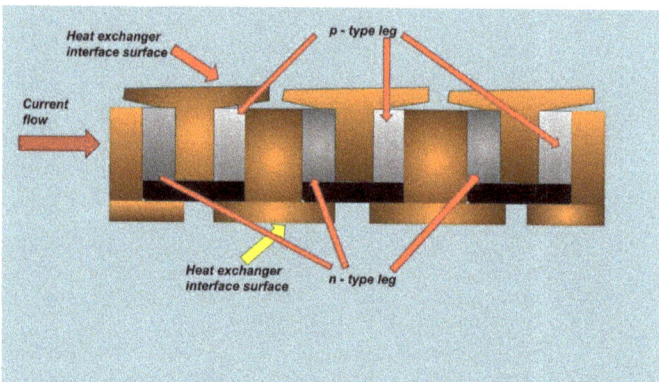

Figure 14. TEG engine system designed and built by Crane [17].

The test carried out for this work consisted of using the information from the experimental measurements to design a conventional TEG; however, unknown data were necessary to correctly develop the design. The source of the known information was the data provided by Crane's paper [17], shown in Table 12.

Table 12. TEG engine data [17].

Material	Bi_2Te_3
T_c	20 °C = 293.15 K
T_h	150 °C = 423.15 K

Figure 15 shows the power curves against the electric current for certain values of T_h, obtained through the computerized model made by authors in [17]. It is observed that they coincide with the experimental curves.

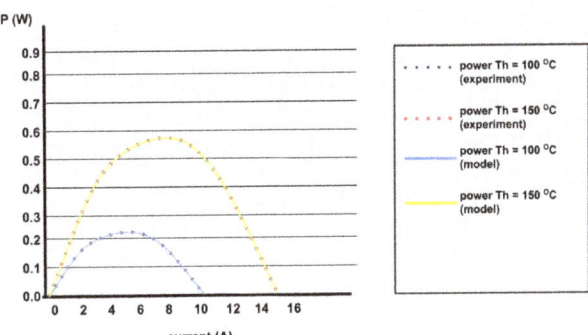

Figure 15. Power and voltage versus current curves to validate the computer model with experimental data taken from Crane et al. [17].

Next, we sought to design a three-couple *TEG* engine system applying the reduced variables methodology with supervised machine learning, to obtain a graph like the one shown in Figure 15. In addition to the data shown in Table 12, it was required to know the properties and dimensions A_p, A_n, l_p, l_n, ρ_p, ρ_n, A_{metal}, l_{metal}, and ρ_{metal} not provided by [17], so it was necessary to calculate them. For this purpose, a first attempt was made, in which the Seebeck coefficient and electrical resistivity data from Table 3 were used to obtain the graph of power at temperature $T_h = 150\,°C = 423.15\,K$, presented as the yellow curve in Figure 15. In this case, the six temperature values that are provided in Table 3 were the real values taken from the measurement and, as can be seen, the extreme values $T_c = 298\,K = 24.85\,°C$ and $T_h = 423\,K = 149.85\,°C$ were only approximate to the values actually required for the design ($T_c = 293.15\,K = 20\,°C$ and $T_h = 423.15\,K = 150\,°C$). With this information, the reduced variables technique was applied to obtain a first design for the three-couple TEG engine, for which the graph of electrical power as a function of electrical current is shown in Figure 16.

A comparison between Figure 16 and the yellow curve in Figure 15 shows that the design obtained in this first attempt deviated from Crane's model by approximately 50%, which is a very high margin. It was therefore necessary to adjust the six temperature values from Table 3 to the correct range of $T_c = 298\,K = 24.85\,°C$ and $T_h = 423\,K = 149.85\,°C$. The following table shows the adjustment.

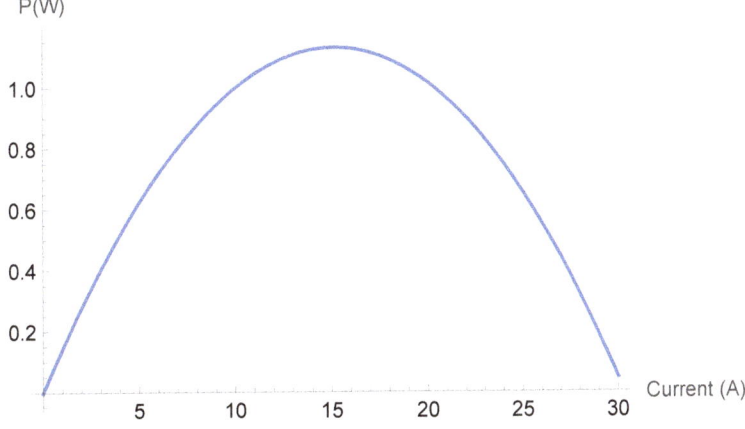

Figure 16. Power vs. current three-couple *TEG* engine at $T_h = 150\,°C$.

The values $u\kappa_n$, $u\kappa_p$, α_n, α_p, ρ_n, and ρ_p (Tables 2 and 3) were used as training data for the supervised machine learning code implemented in this work and thus generated appropriate data from these properties for the temperature values from Table 13. The results are shown in the following Table 14.

Table 13. Adjusted temperatures for the design of the TEG engine.

T (K)
$T_0 = 293.15 = 20\ °C$
$T_1 = 319.15$
$T_2 = 345.15$
$T_3 = 371.15$
$T_4 = 397.15$
$T_5 = 423.15 = 150\ °C$

Table 14. Values Bi_2Te_3 thermoelectric properties obtained by supervised machine learning model.

T (K)	$u_p\kappa_p dT$ (A/cm)	α_p (µV/K)	ρ_p (10^{-3} Ωcm)
$T_0 = 293.15 = 20\ °C$	0.8278	193	1.3975
$T_1 = 319.15$	0.8278	193	1.3976
$T_2 = 345.15$	0.8278	193	1.3976
$T_3 = 371.15$	0.85345	194	1.3977
$T_4 = 397.15$	0.85345	194	1.3978
$T_5 = 423.15 = 150\ °C$	0.85345	193.5	1.3979
T (K)	$u_n\kappa_n dT$ (A/cm)	α_n (µV/K)	ρ_n (10^{-3} Ωcm)
$T_0 = 293.15 = 20\ °C$	−0.5507	−209	2.7573
$T_1 = 319.15$	−0.5507	−209	2.7574
$T_2 = 345.15$	−0.5508	−210	2.7574
$T_3 = 371.15$	−0.5508	−210	2.7575
$T_4 = 397.15$	−0.5508	−187	2.7575
$T_5 = 423.15 = 150\ °C$	−0.5509	−171	2.7576

Using the data from column two and applying the methodology with the set of Equations (2)–(12), the geometric parameters of the couples of the TEG engine were obtained, see Table 15.

Table 15. Geometric parameters obtained for the design of the TEG engine of Bi_2Te_3.

Parameter	Numerical Value
$l_n = l_p$	1.85 mm
A_p	1.73 mm^2
A_n	2.67 mm^2

Subsequently, with the data from the third and fourth columns, the averages of the quantities α_n, α_p, ρ_n, and ρ_p in the temperature range ($T_c = 293.15$ K = 20 °C and $T_h = 423.15$ K = 150 °C) were obtained. The results are shown in Table 16.

Table 16. Averaged Seebeck coefficient and electrical resistivity, TEG engine design.

Property	Averaged Value
$\overline{\alpha_n}$	193.499 (µV/K)
$\overline{\alpha_p}$	−201.111 (µV/K)
$\overline{\rho_n}$	2.75749 ρ_p (10^{-3} Ωcm)
$\overline{\rho_p}$	1.39773 ρ_p (10^{-3} Ωcm)

The data obtained for the legs of the *TEG* engine (Tables 15 and 16), as well as the data of the metallic bridge (Table 10), were used in combination with the following equation for the power produced by the *TEG* engine.

$$P_{TEG-engine} = n\left[(\alpha_p - \alpha_n)(T_h - T_c)I - \left(\frac{\rho_n l_n}{A_n} + \frac{\rho_p l_p}{A_p} + \frac{\rho_{metal} l_{metal}}{A_{metal}}\right)I^2\right] \quad (20)$$

where $P_{TEG-engine}$ is the power produced by the TEG engine; n is the number thermocouples, which in this case is $n = 3$; and I is the electric current (which is the independent variable and is measured in amperes). The other quantities that appear in Equation (20) have already been indicated above. The graph of the electrical power produced by the three-couple *TEG* engine is shown in Figure 17.

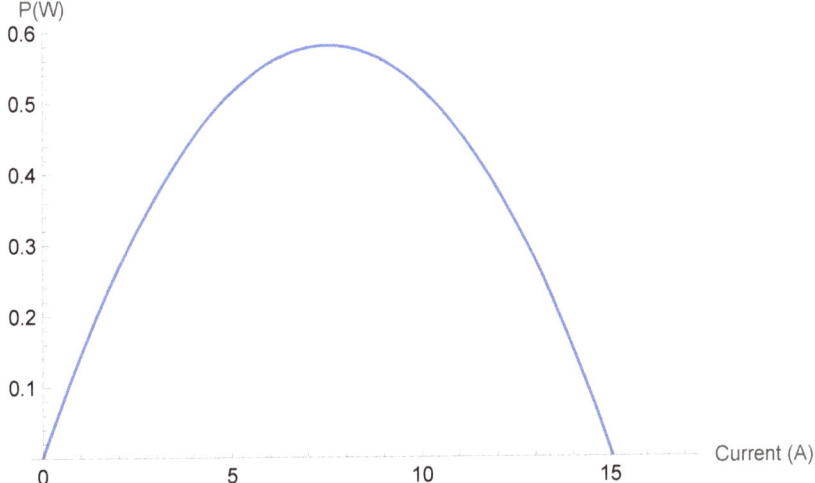

Figure 17. Power vs. current, three-couple *TEG* engine at T_h = 150 °C, applying a supervised machine learning model.

In Figure 17, it can be seen that the graph obtained adjusted well to the original Crane curve at T_h = 150 °C (curve in yellow); see Figure 14. It can be seen that the same maximum value was reached below 0.6 W around 8 A in a range of 0–15 A. Therefore, it is evident that the methodology (which combined the reduced variables method with supervised machine learning) proposed in this work managed to reproduce the three-couple *TEG* engine model and the experiment of Crane et al. [17].

7. Conclusions

This work has shown that the fusion of supervised machine learning with the reduced variables technique can be a useful tool for designing *TEGs* and adjusting them to the conditions imposed by the operating environment of the system, specifically when facing the challenge of knowing a reduced set of measured values. Thanks to its ability to calculate geometric and prediction parameters, it is possible to

(a) Approximate values of thermoelectric properties (α, ρ, κ) for any temperature value;

(b) Generate new data from few experimental values, even when it is not possible to perform a new experiment;

(c) Design *TEGs* for any range of temperature; if the value of a thermoelectric property for a specific temperature is not known experimentally, it is possible to predict it with (*MLS*) (Figures 2 and 5) and use it for design calculations;

(d) Analyze power simultaneously with respect to various parameters for any temperature value and determine optimization conditions.

The previous characteristics were verified through the design and analysis of the power of the conventional TEG (Section 3.1), segmented TEG (Section 4.1), and TEG engine (Section 6) systems.

When conducting the power analysis, one of the outstanding findings was the result shown in Figure 11, indicating that the curve that corresponded to the highest temperature intersected with the curves of the other temperatures. Each intersection point specifically corresponds to the maximum performance of a segment. These intersection points occur very close to the value (A_p/A_n) in which the power of the segmented TEG is maximized.

As shown throughout this work, the proposed methodology was translated into a code in mathematics; its usability is now evaluated in terms of the following ten techno-economic aspects:

(1) The calculation scheme will be updated soon, to introduce a procedure based on heat transfer and to consider the physical aspects of the heat source;

(2) From an economic point of view, although Wolfram is a licensed software, it currently allows the user a free basic plan account, in which the code notebook could be published and shared with those interested in designing thermoelectric systems with this methodology;

(3) The code automatically selects the appropriate prediction method according to the training data. For the study of conventional and segmented systems, the methods that the code selected to predict the values were linear regression, decision tree, and first neighbors;

(4) The methodology could be transferred in code to another type of software that is freely available; for example, it could be implemented in python. In that case, there would be the advantage of being able to modify the prediction method and make more robust codes that can be better adjusted to the training data;

(5) This methodology may allow for the design of $TEGs$, taking advantage of the results of experimental measurements reported in various papers. This aspect is very useful for researchers who want to design $TEGs$ and who do not have a laboratory, the equipment to develop experiments, or specialized software for design and simulation;

(6) So far, the code has been tested with the thermoelectric materials Bi_2Te_3, Zn_4Sb_3, and $CoSb_3$. Currently, there are new materials, for example, organic materials or new alloys; thus, tests must be carried out using the properties of these new materials, to adapt the code to the needs of new TEG applications.

(7) In addition to what is mentioned in point (5), the designed code does not require high-end computing equipment compared to specialized design software. In the case of the calculations developed in this work, an AMD Ryzen 3 processor with Radeon Vega Mobile Gfx 2.60 GHz, with 8 GB RAM, was used;

(8) The methodology has only been applied for constant cross-sectional areas with a quadrangular geometry. The code should be extended to include the design of $TEGs$ with cross-sectional areas with geometries other than quadrangular, for example, circular. Variable cross-sectional areas regarding leg length could also be included. This could be achieved by reformulating Equations (16) and (17) in terms of A_n and A_p;

(9) The code does not send any warning in the event that the user makes some type of error when capturing the information for the design; this still depends on the skills and knowledge of the user regarding the specifications of the system to be designed, but it is intended that soon some kind of table with reference values will be added to act as a design guide;

(10) It would be very useful to link this code with an interface that collects data in real time from experiments carried out in various laboratories around the world. This would allow various thermoelectric material research groups to make a quick evaluation of the possibility of developing new $TEGs$ devices.

Artificial Intelligence Applied in Thermoelectrics

Supervised machine learning (which is an area of AI) has been used to predict values of thermoelectric properties and the electrical power produced; the input values are the operating temperature and the space between the legs. The application of this powerful computational tool is novel in the field of thermoelectric devices, as seen in [18]. Artificial neural networks were applied to model a thermoelectric generator's maximum energy generation and efficiency. The authors concluded that neural networks demonstrated an extremely high prediction accuracy, greater than 98%, and they can operate under a constant temperature difference and heat flux. The physical model considered contact resistance, electrical, surface heat transfer, and other thermoelectric effects.

Furthermore, Chika Maduabuchi's paper [19] presented the first AI-enabled optimization of a TEG performed using deep neural networks (DNN). The effects of strategic parameters on TEG power output, efficiency, and thermal stress performance were investigated. The parameters were hot and cold junction temperatures, heat transfer coefficients, incident heat flux, external load resistance, span height TE, area, and area ratio.

Author Contributions: Conceptualization, methodology, software, validation, formal analysis, investigation, writing—original draft preparation, writing—review and editing: A.V.-A., M.A.O.-R. and A.A.A.-V. All authors have read and agreed to the published version of the manuscript.

Funding: This research was funded by the Consejo Nacional de Ciencia y Tecnología (CONACYT), grant number 444915, Instituto Politecnico Nacional, grant number 20230037, Consejo de Ciencia y Tecnología del Estado de Tabasco (CCYTET) and Universidad Politécnica del Golfo de México, grant number PRODECTI-2022-01/079.

Institutional Review Board Statement: Not applicable for studies not involving humans or animals.

Data Availability Statement: Not applicable.

Acknowledgments: Acknowledgments are given to the Instituto Politécnico Nacional (IPN) and Universidad Politécnica del Golfo de Mexico (UPGM) for their administrative and technical support.

Conflicts of Interest: The authors declare no conflict of interest.

Sample Availability: Samples of the compounds are available from the authors.

Appendix A

Finally, it is worth mentioning that the algorithm is currently in training to advance the knowledge of the value of maximum power when introducing any parameter value (H, A_p/A_n, l, l_{n2}/l_{n1}, or l_{p2}/l_{p1}) and any operating temperature value. Figures A1 and A2 show the first advance obtained by training the algorithm by providing maximum power values at temperatures 298, 323, and 348 K with $H = 0.25$ and maximum power values at temperatures 373, 398, and 418 K with $H = 0.5$. After executing the algorithm, a test was conducted, introducing the values of 298, 323, and 348 K with $H = 0.5$ and 373, 398, and 418 K with $H = 0.25$ in the code. Then, observing Figure A1 for the conventional case, from left to right, the predicted results of maximum power when performing the test are observed first, and then the values that were previously known from the spreadsheet are observed. Similarly, this is evident for the case of the segmented system (Figure A2). It can be noted for both systems, that in this first training, the results obtained with the algorithm maintain an acceptable approximation with the spreadsheet results, which were obtained using experimental measurements of the thermoelectric properties reported in the literature.

Predictions generated by the training code	Power results for temperature values (298 K, 323 K, 348 K) with H = 0.5. (373 K, 398 K, 418 K) with H = 0.25			
	Output power (W)	Temperature (K)	Output power (W)	Temperature (K)
In[2]:= p[{298, "0.5"}] Out[2]= 0.0299054	0.030629132	298	0.02671527	373
	0.030490876	300.5	0.0266116	375.5
In[3]:= p[{323, "0.5"}]	0.030354111	303	0.02650874	378
Out[3]= 0.0289807	0.03021875	305.5	0.02649118	380.5
	0.030084335	308	0.02638969	383
In[4]:= p[{348, "0.5"}]	0.029951322	310.5	0.02628898	385.5
Out[4]= 0.0280559	0.029819548	313	0.02618908	388
	0.029688964	315.5		
	0.029304409	323	0.02610467	390.5
In[5]:= p[{373, "0.25"}]	0.029178359	325.5	0.02606458	393
Out[5]= 0.0271715	0.029053742	328	0.02596653	395.5
	0.028930141	330.5		
	0.028807477	333	0.02586921	398
In[6]:= p[{398, "0.25"}]	0.028686163	335.5	0.02577267	400.5
Out[6]= 0.0262468	0.028565625	338	0.02575883	403
	0.028446293	340.5	0.0255874	405.5
	0.028328201	343	0.02549299	408
In[7]:= p[{418, "0.25"}]	0.028247464	345.5	0.02539954	410.5
Out[7]= 0.025507	0.028094593	348	0.02530655	413
			0.02521441	415.5
			0.02512318	418

Figure A1. From left to right, the maximum power values predicted using the algorithm are shown first and then the power values calculated in a spreadsheet using experimental data (yellow color). Conventional thermoelectric generator.

The first results allowed us to determine that it is possible to obtain an algorithm for designing conventional and segmented thermocouples based on the reduced variables approach fused into a supervised machine learning calculation model trained for various thermoelectric materials. The reduced variables technique helps obtain the dimensions of the generator, cross-sectional area, and length of the legs. However, there is the possibility that the values of the thermoelectric properties are only known for certain temperatures. This is a situation that could arise for a researcher who does not have the equipment to carry out experimental measurements. One solution is to use a dataset obtained from the literature to generate a more extensive dataset, applying a supervised machine learning resource. In this work, this was helpful, because it allowed us to generate the values of the thermoelectric properties for any temperature and then calculate the corresponding maximum electrical power.

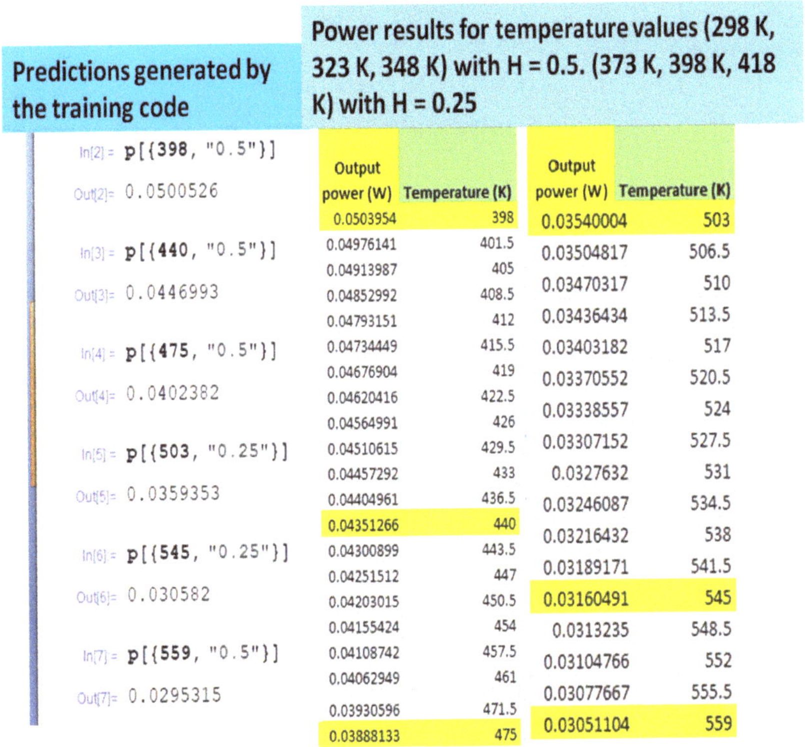

Predictions generated by the training code	Power results for temperature values (298 K, 323 K, 348 K) with H = 0.5. (373 K, 398 K, 418 K) with H = 0.25			
	Output power (W)	Temperature (K)	Output power (W)	Temperature (K)
In[2]:= p[{398, "0.5"}]	0.0503954	398	0.03540004	503
Out[2]= 0.0500526	0.04976141	401.5	0.03504817	506.5
In[3]:= p[{440, "0.5"}]	0.04913987	405	0.03470317	510
Out[3]= 0.0446993	0.04852992	408.5	0.03436434	513.5
	0.04793151	412	0.03403182	517
In[4]:= p[{475, "0.5"}]	0.04734449	415.5	0.03370552	520.5
	0.04676904	419	0.03338557	524
Out[4]= 0.0402382	0.04620416	422.5	0.03307152	527.5
	0.04564991	426	0.0327632	531
In[5]:= p[{503, "0.25"}]	0.04510615	429.5	0.03246087	534.5
Out[5]= 0.0359353	0.04457292	433	0.03216432	538
	0.04404961	436.5	0.03189171	541.5
	0.04351266	440		
In[6]:= p[{545, "0.25"}]	0.04300899	443.5	0.03160491	545
	0.04251512	447	0.0313235	548.5
Out[6]= 0.030582	0.04203015	450.5	0.03104766	552
	0.04155424	454	0.03077667	555.5
In[7]:= p[{559, "0.5"}]	0.04108742	457.5	0.03051104	559
	0.04062949	461		
Out[7]= 0.0295315	0.03930596	471.5		
	0.03888133	475		

Figure A2. From left to right, the maximum power values predicted using the algorithm are shown first and then the power values calculated in a spreadsheet using experimental data (yellow color). Segmented thermoelectric generator.

Appendix B. Details about the Supervised Machine Learning Algorithm Applied in This Work

The algorithm was developed with the Wolfram Mathematica software, and its basic operation consists of receiving a set of values (experimental data); the predictor function was used, which analyzes the data and automatically selects the prediction method that best fits the training data. Part of the experimental data was used for training the algorithm, and the other data was used for verification. The model was compared with the results of the work of Mamoozadeh et al. [1], in which they applied a mathematical–numerical model to optimize the cross-sectional area and length of thermoelectric legs to maximize power and conversion efficiency. Figure A3 shows a graph generated with our model. Figure A4 belongs to the abovementioned work, where the blue curve is the power.

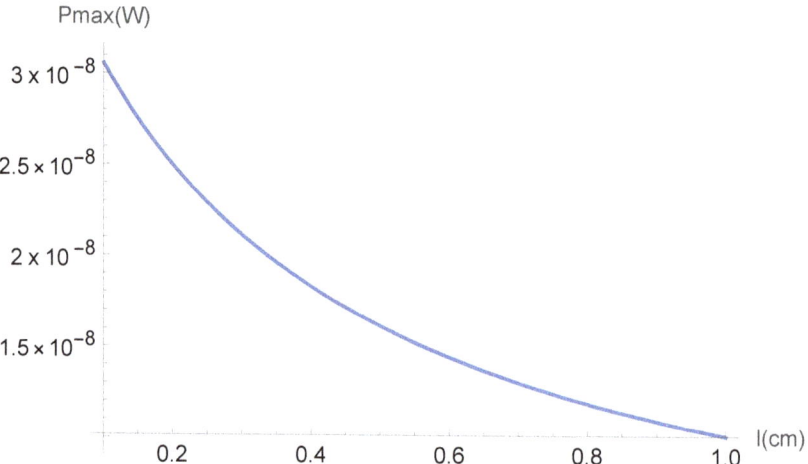

Figure A3. Power curve as a function of the length of the thermoelectric leg, obtained using the algorithm developed in this work.

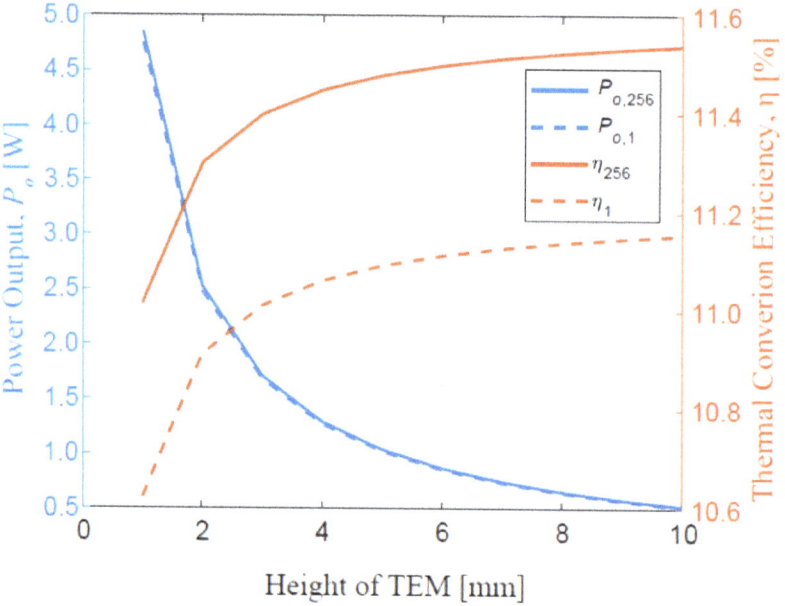

Figure A4. Power curve as a function of the length of the thermoelectric leg, in blue, obtained by [1].

Although the relationship between thermoelectric properties and temperature is not linear, the implemented algorithm works using a predictive function, which improves the degree of approximation in each iteration. The algorithm was tested using a group of certified values of the Seebeck coefficient (reference [20], Figure A5). It can be seen that the algorithm managed to predict the Seebeck coefficient values for temperatures of 350, 450, 550, 650, and 850 K with an acceptable approximation (Figure A6).

Temperature (K)	Certified Values (μV/K)
295	116.25
350	128.87
400	139.88
450	150.46
500	160.59
550	170.29
600	179.54
650	188.36
700	196.74
750	204.68
800	212.18
850	219.24
900	225.87

Figure A5. Certified Seebeck coefficient values obtained from reference [20].

```
In[ ]:= p[350]
        p[450]
        p[550]
        p[650]
        p[850]


Out[ ]= 131.186

Out[ ]= 148.78

Out[ ]= 166.374

Out[ ]= 183.969

Out[ ]= 219.158
```

Figure A6. Seebeck coefficient values generated by the algorithm developed in this work.

References

1. Mamoozadeh, A.K.; Wielgosz, S.E.; Yu, K.; Drymiotis, F.; Barry, M.M. Optimization of variable cross-sectional area thermoelectric elements through multi-method thermal-electric coupled modeling. In Proceedings of the 5th–6th Thermal and Fluids Engineering Conference (TFEC), Online, 26–28 May 2021.
2. Ge, Y.; He, K.; Xiao, L.; Yuan, W.; Huang, S.-M. Geometric optimization for the thermoelectric generator with variable cross-section legs by coupling finite element method and optimization algorithm. *Renew. Energy* **2022**, *183*, 294–303. [CrossRef]
3. Şişik, B.; LeBlanc, S. The Influence of Leg Shape on Thermoelectric Performance under Constant Temperature and Heat Flux Boundary Conditions. *Front. Mater.* **2020**, *7*, 595955. [CrossRef]
4. Bian, M.; Xu, Z.; Meng, C.; Zhao, H.; Tang, X. Novel geometric design of thermoelectric leg based on 3D printing for radioisotope thermoelectric generator. *Appl. Therm. Eng.* **2022**, *212*, 118514. [CrossRef]
5. Jaziri, N.; Boughamoura, A.; Müller, J.; Mezghani, B.; Tounsi, F.; Ismail, M. A comprehensive review of Thermoelectric Generators: Technologies and common applications. *Energy Rep.* **2020**, *6*, 264–287. [CrossRef]
6. Enescu, D. Thermoelectric Energy Harvesting: Basic Principles and Applications. In *Green Energy Advances*; IntechOpen: London, UK, 2019.
7. Wijesekara, W.; Rosendahl, L. Expanding the reduced-current approach for thermoelectric generators to achieve higher volumetric power density. *Phys. Status Solidi A* **2015**, *212*, 591–599. [CrossRef]
8. Snyder, G.J. Thermoelectric Power Generation: Efficiency and Compatibility. Available online: https://thermoelectrics.matsci.northwestern.edu/publications/pdf/CompatCRC.pdf (accessed on 10 January 2023).
9. Almeida, A.V.; Olivares Robles, M.A. Geometric conditions for minimizing entropy production in thermocouple design. *Results Phys.* **2022**, *41*, 105893. [CrossRef]
10. Snyder, G.J. Thermoelectric Power Generation: Efficiency and Compatibility. In *Chapter 9, Thermoelectrics Handbook, Macro to Nano*; CRC Press: Boca Raton, FL, USA, 2006. [CrossRef]
11. Han, G.; Sun, Y.; Feng, Y.; Lin, G.; Lu, N. Machine Learning Regression Guided Thermoelectric Materials Discovery—A Review. *ES Mater. Manuf.* **2021**, *14*, 20–35. [CrossRef]
12. Mbaye, M.T.; Pradhan, S.K.; Bahoura, M. Data-driven thermoelectric modeling: Current challenges and prospects. *J. Appl. Phys.* **2021**, *130*, 190902. [CrossRef]
13. Iwasaki, Y.; Takeuchi, I.; Stanev, V.; Kusne, A.G.; Ishida, M.; Kirihara, A.; Ihara, K.; Sawada, R.; Terashima, K.; Someya, H. ; et al. Machine-learning guided discovery of a new thermoelectric material. *Sci. Rep.* **2019**, *9*, 2751. [CrossRef] [PubMed]
14. Zhao, X.; Shi, L.; Tian, B.; Li, S.; Liu, S.; Li, J.; Liu, S.; James, T.D.; Chen, Z. Harnessing solar energy for electrocatalytic biorefinery using lignin-derived photothermal materials. *J. Mater. Chem. A* **2023**, *11*, 12308–12314. [CrossRef]
15. Moser, W.; Friedl, G.; Aigenbauer, S.; Heckmann, M.; Hofbauer, H. A Biomass-Fuel based Micro-Scale CHP System with Thermoelectric Generators. In Proceedings of the Central European Biomass Conference, Graz, Austria, 16–19 January 2008. Available online: http://hdl.handle.net/20.500.12708/47274 (accessed on 10 January 2023).
16. Liu, C.; Qu, G.; Shan, B.; Aranda, R.; Chen, N.; Li, H.; Zhou, Z.; Yu, T.; Wang, C.; Mi, J.; et al. Underwater hybrid energy harvesting based on TENG-MTEG for self-powered marine mammal condition monitoring system, Materials Today Sustainability. In Proceedings of the Central European Biomass Conference 2008, Messe Center, Graz, Austria, 16–19 January 2008; Volume 21.
17. Crane, D.T.; LaGrandeur, J.W.; Harris, F.; Bell, L.E. Performance Results of a High-Power-Density Thermoelectric Generator: Beyond the Couple. *J. Electron. Mater.* **2009**, *38*, 1375–1381. [CrossRef]
18. Zhu, Y.; Newbrook, D.W.; Dai, P.; de Groot, C.K.; Huang, R. Artificial neural network enabled accurate geometrical design and optimisation of thermoelectric generator. *Appl. Energy* **2022**, *305*, 117800. [CrossRef]
19. Maduabuchi, C. Thermo-mechanical optimization of thermoelectric generators using deep learning artificial intelligence algorithms fed with verified finite element simulation data. *Appl. Energy* **2022**, *315*, 118943. [CrossRef]
20. National Institute of Standards and Technology. Standard Reference Material 3452, High-Temperature Seebeck Coefficient Standard (295 K to 900 K). Certificate of Analysis, pp. 1–6. David Holbrook, Chief Materials Measurement Science Division; Gaithersburg, MD 20899, Certificate Issue Date: 4 February 2021. Available online: https://www.nist.gov/srm (accessed on 10 January 2023).

Disclaimer/Publisher's Note: The statements, opinions and data contained in all publications are solely those of the individual author(s) and contributor(s) and not of MDPI and/or the editor(s). MDPI and/or the editor(s) disclaim responsibility for any injury to people or property resulting from any ideas, methods, instructions or products referred to in the content.

Article

Analyzing the Performance of Thermoelectric Generators with Inhomogeneous Legs: Coupled Material–Device Modelling for Mg$_2$X-Based TEG Prototypes

Julia Camut [1], Eckhard Müller [1,2] and Johannes de Boor [1,3,*]

1. Department of Thermoelectric Materials and Systems, Institute of Materials Research, German Aerospace Center, 51147 Cologne, Germany
2. Institute of Inorganic and Analytical Chemistry, JLU Giessen, 35390 Giessen, Germany
3. Institute of Technology for Nanostructures (NST) and CENIDE, Faculty of Engineering, University of Duisburg-Essen, 47057 Duisburg, Germany
* Correspondence: johannes.deboor@dlr.de

Citation: Camut, J.; Müller, E.; de Boor, J. Analyzing the Performance of Thermoelectric Generators with Inhomogeneous Legs: Coupled Material–Device Modelling for Mg$_2$X-Based TEG Prototypes. *Energies* 2023, *16*, 3666. https://doi.org/10.3390/en16093666

Academic Editor: Diana Enescu

Received: 6 March 2023
Revised: 6 April 2023
Accepted: 8 April 2023
Published: 24 April 2023

Copyright: © 2023 by the authors. Licensee MDPI, Basel, Switzerland. This article is an open access article distributed under the terms and conditions of the Creative Commons Attribution (CC BY) license (https://creativecommons.org/licenses/by/4.0/).

Abstract: Thermoelectric generators (TEGs) possess the ability to generate electrical power from heat. As TEGs are operated under a thermal gradient, inhomogeneous material properties—either by design or due to inhomogeneous material degradation under thermal load—are commonly found. However, this cannot be addressed using standard approaches for performance analysis of TEGs in which spatially homogeneous materials are assumed. Therefore, an innovative method of analysis, which can incorporate inhomogeneous material properties, is presented in this study. This is crucial to understand the measured performance parameters of TEGs and, from this, develop means to improve their longevity. The analysis combines experimental profiling of inhomogeneous material properties, modelling of the material properties using a single parabolic band model, and calculation of device properties using the established Constant Property Model. We compare modeling results assuming homogeneous and inhomogeneous properties to the measurement results of an Mg$_2$(Si,Sn)-based TEG prototype. We find that relevant discrepancies lie in the effective temperature difference across the TE leg, which decreases by ~10%, and in the difference between measured and calculated heat flow, which increases from 2–15% to 9–16% when considering the inhomogeneous material. The approach confirms additional resistances in the TEG as the main performance loss mechanism and allows the accurate calculation of the impact of different improvements on the TEG's performance.

Keywords: thermoelectrics; performance modelling; material modelling; TEG characterization; single parabolic band model; constant property model; inhomogeneous material; performance analysis

1. Introduction

Thermoelectric (TE) materials can convert heat flow into electrical power. They have gained a lot of interest over the past decades as a green source of electrical energy [1] since 60% of the primary energy is lost as waste heat [2]. TE devices have the advantages of being lossless scalable, and negligible maintenance requirements due to their lack of moving parts. This has made them a reliable energy source in demanding fields, such as the aerospace industry, in which RTGs (Radioisotope Thermoelectric Generators) were used in several space missions [3]. Terrestrial applications are also being considered and researched, such as in the automotive industry and industrial processes [3–5], wearable medical devices [6–8], mobile storage of pharmaceuticals, and electronic devices [3].

A thermoelectric generator (TEG) is a device in which n- and p-type TE legs are connected electrically in series and thermally in parallel to generate an electrical current from a heat flow. Over the last decades, the optimization of the TE properties of various material classes has been the main focus of research as the first, very challenging step in the development chain of a TEG [9–18]. As a consequence, research on contacting techniques and on TE module building remained relatively scarce, and only TEGs based on $(Bi,Sb)_2(Te,Se)_3$ and PbTe/TAGS have reached commercial maturity. For many other promising material classes such as half-Heusler materials, Skutterudites, PbTe, and Zintl-phases, prototypes have been demonstrated [19–23]; partially, with limited stability due to an early development stage.

Another promising material class is solid solutions from Mg_2Si-Mg_2Sn, due to the high performance of the n-type material, the recent improvement of the p-type material, and the abundance and non-toxicity of its components [12,13,24–26]. It is also lightweight; which, is advantageous for mobile applications. For this material class, few prototypes have been reported recently [27,28], and we have performed the first efficiency measurement on an $Mg_2(Si,Sn)$-based TEG, demonstrating 3.6% conversion efficiency for a hot side temperature of 400 °C and a cold side temperature of 25 °C [29].

Locally resolved measurements showed a spatially varying Seebeck coefficient, indicating an inhomogeneous carrier concentration in the initially homogeneous material. As the device properties depend on the figure of merit $zT = \frac{\alpha^2 \sigma}{\kappa} T$ of the employed material and the material properties α (Seebeck coefficient), σ (electrical conductivity), and κ (thermal conductivity), all depend on the carrier concentration that has a direct impact on the device's performance. In fact, spatially inhomogeneous materials are quite a common feature in TEGs. With respect to our prototype, $Mg_2(Si,Sn)$ was previously shown to be sensitive to Mg evaporation at expected working temperatures [30,31] and in addition to interdiffusion with foreign elements, such as the metallization layers which are included in a TEG design [32,33]. Both mechanisms usually lead to spatially inhomogeneous changes in the carrier concentration due to changes in the intrinsic or extrinsic defect densities. Other material systems show similar chemical or thermal instabilities, leading to inhomogeneity. Such as: Mg_3Sb_2, which is also sensitive to Mg loss [34]; $CoSb_3$ and half-Heusler compounds, which are sensitive to Sb loss [35]; as well as $(Bi,Sb)_2(Te,Se)_3$, which is sensitive to corrosion [36]. Generally, due to high temperatures (processing, application) and the proximity of multiple chemical elements, doping defects can form in the TE materials, which alter their properties [32]. On the other hand, local variation of properties can also be intended and designed, as thermoelectric material properties are generally quite strongly temperature-dependent; in addition, strategies such as segmentation [25,37] or grading are employed for performance optimization [38].

Those local changes in TE properties can be challenging when it comes to TEG modeling, used to predict the performance of a TEG, and in providing a reference to evaluate and understand measured data. The Constant Property Model (CPM) is generally used in TEG calculations for its simplicity and ease [25,39–41]. For real materials, averages corresponding to the relevant temperature range/profile are employed, resulting in a relatively accurate prediction of TEG performance [32,42]. However, these averages are obtained from temperature-dependent data; which, are usually obtained by integral bulk measurements on material samples before device manufacturing, and are therefore only available for homogeneous materials. On the other hand, locally resolved properties of inhomogeneous samples are measured typically only at room temperature [43,44]. To be able to model inhomogeneous materials in general, but also to calculate correct average values for the CPM, there is a need for a material model that is both spatially and temperature dependent. The performance of inhomogeneous thermoelectric materials was already modeled using the Effective Media Theory [45–47]; however, those works are generally applied to (intended) nanostructured thermoelectric materials, i.e., typically a composite structure made of particles of a first material embedded in a matrix made of a second material. This is a different case than that addressed here, as these materials are homogeneous on a mesoscale/macroscale.

In this study, an innovative analytical method is developed combining experimental carrier concentration profiling and the Single Parabolic Band (SPB) model combined with the CPM/continuum theory: the CPIM (Constant Property (model) for Inhomogeneous Materials). This method is able to capture the inhomogeneity of thermoelectric materials on a macroscopic/mesoscopic scale and is employed exemplarily to the measurement results of an Mg$_2$(Si,Sn)-based TEG. The predictions considering the actual inhomogeneous properties of the n-type legs are compared to those assuming a homogeneous material and the experimental data. When considering the inhomogeneous material, the deviation between calculations and measurement of the heat flow increases above the measurement uncertainty, and higher thermal losses through the TEG/heat exchanger coupling are observed. However, for both conditions, the high difference observed between calculated and measured electrical resistance likely indicates crack formation; which, is a commonly observed degradation mechanism in TEGs.

2. Methods

Experimental:

This work deals with the same TEG as reported in [29], where the building process and characterization methods are described. In this study, the solid solution Mg$_2$Si$_{0.3}$Sn$_{0.7}$ was chosen as the chemical composition for both p- and n-type materials of the TEG. Both materials have similar mechanical properties [48] and have a reported figure of merit among the best in the mid- to high-temperature range. Pictures and dimensions are reported in [29], while a schematic of the prototype is shown in Figure 1. The powder used to sinter the n-type material was synthesized and compacted, such as reported in [24]; the resulting properties are shown in Figure 2. The p-type powder is synthesized similarly, as described in [32].

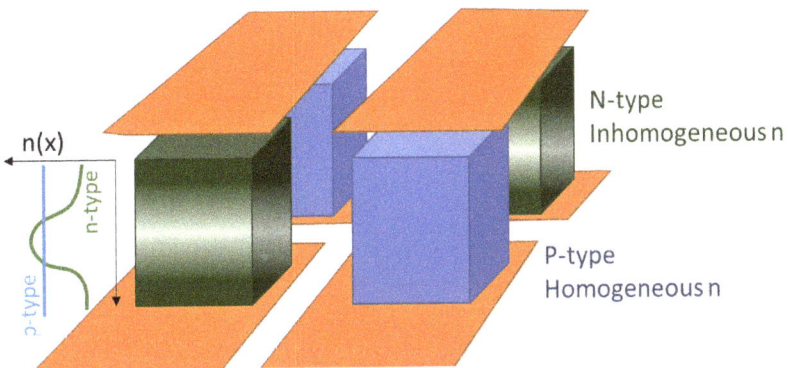

Figure 1. Sketch of a 2 × 2 TEG; where, the color gradient in the n-type legs indicates the inhomogeneity in the carrier concentration. The top bridges are not directly connected to the legs on the sketch to have a better view of the full length of the legs; in reality, they are bonded with a metallization layer.

The Seebeck coefficient and electrical conductivity were measured using an in-house device with a four-probe technique (HTSσ) [44,49]. The thermal conductivity κ was calculated from $\kappa = D\rho C_p$, where D, ρ, and C_p are the sample thermal diffusivity, density, and heat capacity depending on the composition at constant pressure, respectively. D was measured by a laser flash technique with a NETZSCH LFA 427 apparatus or with an XFA 467HT Hyperflash apparatus; ρ was measured using Archimedes' method. $C_p = C_v^{DP} + \frac{9E_T^2 T}{\beta_T \rho}$, where C_v^{DP} is the Dulong-Petit limit; E_T and β_T, respectively, are the linear coefficient of thermal expansion and isothermal compressibility dependent on composition [50]. The values for Mg$_2$Si$_{0.3}$Sn$_{0.7}$ are $2 \cdot 10^{-5}$ K^{-1} and $2.07 \cdot 10^{-11}$ Pa^{-1}, respectively [43]. The measurement uncertainties for α, σ, and κ are ±5%, ±5%, and ±8%, respectively, based on a comparison with the NIST low-temperature standard for the Seebeck coefficient [51] and internal reference measurements on a high-temperature standard [52]. Our estimates are comparable to those obtained in an international Round-robin test.

The metallization of the pellets was done using Al foils with Zn coatings, such as those reported in [53]. The resulting p-type legs are homogeneous while the n-type legs show a property gradient, presumably due to Zn diffusion.

CPIM (Constant Property model for Inhomogeneous Materials):

The CPIM relies on the application of the CPM (Constant Property Model) to inhomogeneous materials. The inhomogeneity is implemented in the model by experimentally obtaining a spatial distribution of the carrier concentration in the TE material, using a Seebeck coefficient microprobe and the SPB (Single Parabolic Band) model to link both quantities. The SPB is then used to obtain a spatial temperature-dependent distribution of all relevant quantities for the calculation of the average values used in the CPM. Each component of the model is adequately detailed below.

SPB model:

To capture the inhomogeneous properties of the n-type legs and be able to predict the properties at temperatures higher than room temperature, a single parabolic band model is employed. An SPB model allows us to calculate the macroscopic transport n-type properties based on a few material parameters: the reduced chemical potential ($\eta = \frac{E_F}{k_B T}$), where E_F is the Fermi energy and k_B is Boltzmann's constant); the mobility parameters for acoustic phonon scattering (AP) and alloy scattering (AS) ($\mu_{0,AP}$ and $\mu_{0,AS}$, respectively); and the density of states effective mass (m_D^*) [50,54]. In this work, the lattice thermal conductivity (κ_{lat}) is also used as an input parameter. The transport properties are obtained using the following equations; which, are given here in the specific case corresponding to AP and AS as relevant scattering mechanisms.

$$\alpha = \frac{k_B}{e}\left(\frac{2F_1}{F_0} - \eta\right) \quad (1)$$

$$n = 4\pi\left(\frac{2m_D^* k_B T}{h^2}\right)^{1.5} F_{\frac{1}{2}}(\eta) \quad (2)$$

$$\mu_{AP} = \mu_{0,AP} \cdot \psi(\eta) = \frac{\sqrt{8\pi} e \hbar^4 \rho v_l^2}{3 E_{Def}^2 m_s^{2.5} (k_B T)^{1.5}} \psi(\eta) \quad (3)$$

$$\mu_{AS} = \mu_{0,AS} \cdot \psi(\eta) = \frac{64 e \hbar^4 N_0}{9(2\pi)^{1.5} x(1-x) E_{AS}^2 m_s^{2.5} (k_B T)^{0.5}} \psi(\eta) \quad (4)$$

$$\frac{1}{\mu} = \frac{1}{\mu_{AP}} + \frac{1}{\mu_{AS}} \quad (5)$$

$$\sigma = \mu e n \quad (6)$$

$$\kappa = \kappa_{lat} + \kappa_e = \kappa_{lat} + L\sigma T \quad (7)$$

where $\hbar$ is the reduced Planck constant, $F_i(\eta)$ the Fermi integral of order i, and x is the alloy atomic composition in Sn, such as $Mg_2Si_{1-x}Sn_x$ ($x = 0.7$ in this work), $\psi(\eta) = \frac{3\sqrt{\pi}}{16}\frac{F_{-0.5}(\eta)}{F_0(\eta)}$ and $L = \left(\frac{k_B}{e}\right)^2 \frac{3F_0(\eta)F_2(\eta) - 4F_1^2}{F_0(\eta)^2}$.

The other parameters are described in Table 1. These parameters were obtained from the literature for samples whose properties match ours [12], and are therefore applicable.

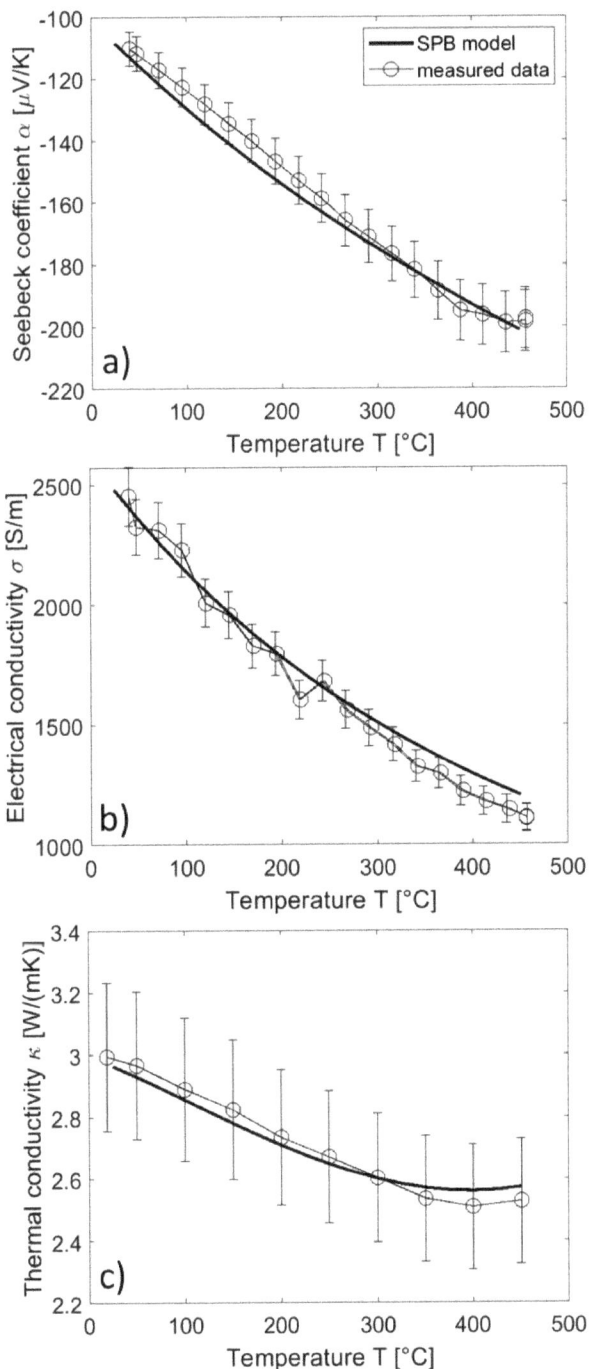

Figure 2. Comparison between the SPB modeled data (for n = 2.3·10^{26} m^{-3}) based on the microscopic parameters in Table 1 and measured data: (**a**) Seebeck coefficient, (**b**) electrical conductivity, and (**c**) thermal conductivity.

Table 1. Parameters used for the Single Parabolic Band model calculations of $Mg_2Si_{0.3}Sn_{0.7}$.

Parameter (Units)	Symbol	Value (SI)	Reference
Average density of states effective mass	m_D^*	$2.1\,m_0$	[50]
Band degeneracy	N_v	6	[31]
Single band mass	m_s	$m_D^*/N_v^{2/3}$	-
Theoretical mass density (g/cm^3)	ρ_D	3.117	[50]
Longitudinal speed of sound (m/s)	v_l	5290	linear with x, [50]
AP deformation potential constant (eV)	E_{Def}	9.8	[55]
Alloy scattering potential (eV)	E_{AS}	0.5	[31,55]
Number of atoms per unit volume (m^{-3})	N_0	$4.105 \cdot 10^{28}$	linear with x, [31]

κ_{lat} is obtained from measured experimental data of our as-sintered n-type material: $\kappa_{lat} = \kappa_{exp} - L\sigma_{exp}T \approx \kappa_{exp} - \left[1.5 + \exp\left(-\frac{|\alpha_{exp}|}{116}\right)\right]\sigma_{exp}T$, where κ_{exp}, σ_{exp}, α_{exp} denote the bulk measured values, corresponding to the homogeneous material, before contacting and device making [11]. The obtained data is fitted with a third-order polynomial $\kappa_{lat} = 1.63\left[\frac{W}{mK}\right] - 2.21 \cdot 10^{-3}\left[\frac{W}{mK^2}\right] \cdot (T-273) + 1.21 \cdot 10^{-6}\left[\frac{W}{mK^3}\right] \cdot (T-273)^2 + 3.09 \cdot 10^{-9}\left[\frac{W}{mK^4}\right] \cdot (T-273)^3$ with T in K, and this fit equation is used to calculate the temperature- and position-dependent total thermal conductivity using the SPB model.

The comparison between bulk measurements on homogeneous samples corresponding to the n-type material employed for the TEG and the related SPB model results (for $n = 2.3 \cdot 10^{26}$ m^{-3}, obtained with the experimental room temperature Seebeck coefficient and Equation (2)) and measured data, is shown in Figure 2.

It can be seen that SPB represents the experimental data well and captures the temperature dependence of the transport properties. $\alpha(T)$ starts to show a different temperature dependence only for $T \geq 400\,°C$ due to the excitation of minority carriers; which, is beyond the analyzed temperature range. Therefore, SPB is fully valid for the initial, homogeneous material.

Obtaining the spatial carrier concentration profile:

The spatial variation of the Seebeck coefficient is obtained using an in-house-built device called the Potential & Seebeck Microprobe (PSM) [56,57]. This device locally measures the Seebeck coefficient and the voltage along a conductive sample; which, in the case of a TE leg, allows us to calculate the electrical contact resistance and map the Seebeck coefficient. Exemplary line scans, obtained in the PSM for the legs prior to TEG making, are shown in Figure 3. In [29], we measured an n-type leg of the TEG post cycling, and little difference was observed in the PSM Seebeck coefficient profile, showing a relatively stable behavior of the legs through TEG measurement. Therefore, pre-cycling profiles can be used for the SPB calculations in this work.

The Seebeck coefficient values obtained with the PSM are systematically underestimated, and not as accurate as those obtained under integral measurement conditions using the HTSσ device. This is due to the temperature difference between the effective position of the thermocouple junction and the point where the thermovoltage is measured [56]; which, leads to an empirically determined deviation between 10% and 20% of the measured Seebeck values in the PSM. This deviation depends on sample properties and tip wear, and was found for the range of thermal conductivities of TE materials (2 to 6.5 W/(mK), respectively, for Bi_2Te_3 and $FeSi_2$); in which, the range our inhomogeneous material falls. This effect is also known as the cold finger effect [43,51]. Note though, that the effect of statistical noise can be reduced by averaging.

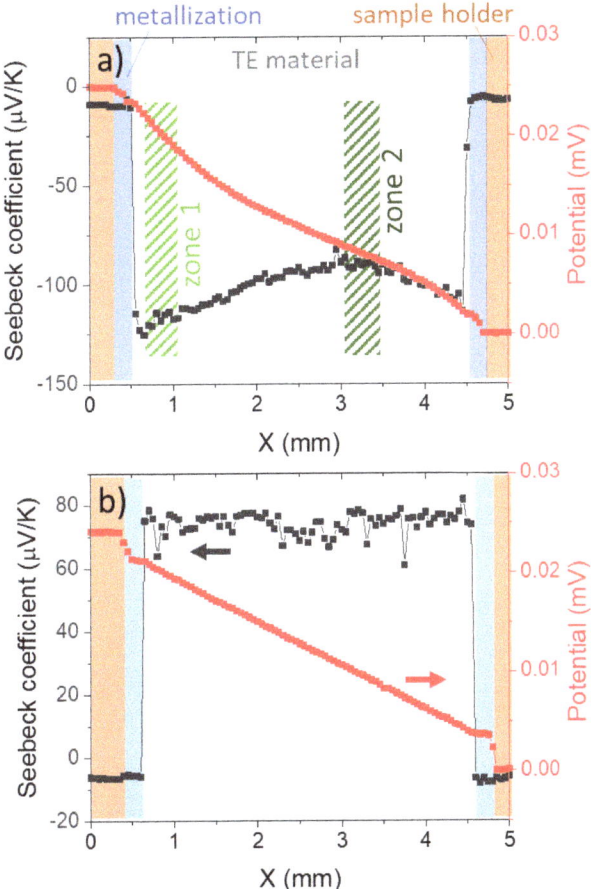

Figure 3. Seebeck coefficient and electrical potential line scans of legs after contacting before TEG joining: (**a**) n-type and (**b**) p-type. In (**a**), it can be noticed that the slope of the potential in zone 2 is steeper than within zone 2, where the Seebeck coefficient is smaller.

A spatial Seebeck coefficient profile at room temperature is obtained by scanning an n-type leg with the PSM. The obtained profile $\alpha_{PSM}(x)$ is converted into a "true" $\alpha(x)$ profile using previously measured Seebeck coefficient values from integral measurement conditions α_{int}. The employed assumptions here are first, that there is a constant relative difference between α_{PSM} and α (which is plausible as the cold finger effect on the PSM measurements leads to a constant relative error) and second, that the carrier concentration in the middle of the sample ($x = x_0$) is unchanged due to the distance to outside metallization layers. This allows us to obtain a corrected Seebeck coefficient profile from $\alpha(x) = \alpha_{PSM}(x) * \alpha_{int}/\alpha_{PSM}(x_0)$, where $\alpha_{PSM}(x_0) = -90$ µV/K and $\alpha_{int} = -109$ µV/K. The deviation between those values is 17%; which, lies within the combined uncertainty for the local Seebeck coefficient by the PSM (10–20%) and that of the integral Seebeck coefficient measurement system (5%).

The assumption that the carrier concentration in the middle of the sample is unchanged is, in principle, an uncertainty of the CPIM. If that is not true, the correction factor and subsequent analysis are flawed. However, for the considered example, this assumption is reasonable since the PSM value at x_0 is close to values measured for as-sintered pellets made with other powder batches of the same composition and similar properties.

As described in Equations (1) and (2), the Seebeck coefficient at room temperature can be linked to the carrier concentration using the SPB model. A carrier concentration profile $n(x)$ can therefore be obtained from the corrected Seebeck coefficient profile.

The temperature function $T(x)$ is obtained along the leg assuming a linear profile between $T_{h,TE}$ and $T_{c,TE}$ (see CPM section). For Mg_2X materials, the linearity of the temperature profile can be assumed in spite of the interplay between Thomson heat, $\kappa(T)$, and Joule heat, as shown by Ponnusamy et al. [42,58].

Constant Property Model (CPM):

The basics of the CPM are given in [39–41] while those applied to the TEG are derived in detail in [29]; however, the most relevant equations are given below and the relevant parameters are reported in Table 2:

$$\Delta T_{TE,0} = \frac{U_{0,m}}{N(\alpha_p - \alpha_n)} \quad (8)$$

$$T_{h,TE,I} = T_{h,m} - 0.5 \times \Delta T_{par,I} \quad (9)$$

$$T_{c,TE,I} = T_{c,m} + 0.5 \times \Delta T_{par,I} \quad (10)$$

$$\frac{Q_{opt,m}}{Q_{0,m}} = \frac{\Delta T_{par,opt}}{\Delta T_{par,0}} = \frac{\Delta T_{m,opt} - \Delta T_{TE,opt}}{\Delta T_{par,0}} \quad (11)$$

$$Q_I = K_{TE}\Delta T_{TE,I} + I \cdot N \cdot (\alpha_p - \alpha_n) T_{h,TE,I} - \frac{1}{2}I^2 R \quad (12)$$

$$I_{opt} = \frac{N(\alpha_p - \alpha_n)\Delta T_{TE,opt}}{2R} \quad (13)$$

$$R = R_{TE} + R_c + R_{Cu} = N\left[\frac{\rho_p L}{A_p} + \frac{\rho_n L}{A_n} + 2r_c\left(\frac{1}{A_p} + \frac{1}{A_n}\right)\right] + R_{Cu} \quad (14)$$

$$K_{TE} = N\left(\frac{\kappa_p A_p}{L} + \frac{\kappa_n A_n}{L}\right) \quad (15)$$

$$P_{max} = N(P_n + P_p) = \frac{(N(\alpha_p - \alpha_n)\Delta T_{TE,opt})^2}{4R} \quad (16)$$

$$\eta_{max} = \frac{T_{h,TE,opt} - T_{c,TE,opt}}{T_{h,TE,opt}} \frac{\sqrt{1+ZT_m} - 1}{\frac{T_{c,TE,opt}}{T_{h,TE,opt}} + \sqrt{1+ZT_m}} \quad (17)$$

$$Z = \frac{N^2(\alpha_p - \alpha_n)^2}{K_{TE} R} \quad (18)$$

where:

Table 2. Parameters of the CPM.

Symbol	Description
m (subscript)	Indicates measured value
U_0	Seebeck voltage
$T_{h,m,I}$, $T_{c,m,I}$	Temperature at the hot, cold block in TEG measurement at current I

Table 2. Cont.

Symbol	Description
$T_{h,TE,I}$, $T_{c,TE,I}$	Temperature at the hot, cold side of the TE legs at current I
$\Delta T_{par,I} = \Delta T_{m,I} - \Delta T_{TE,I}$	Parasitic temperature loss at current I
Q_I	Heat flow at the hot side (Q_{in}) at current I
K_{TE}	Thermal conductance of the TE legs
R_{TE}	Electrical resistance of the TE legs
R_c	Electrical contact resistance
r_c	Electrical contact resistivity
N	Number of leg pairs
L, A	Length, Cross-sectional area of TE element
$R_{Cu} = \sum_i \frac{L_i}{\sigma_{Cu}(T) A_i}$	Resistance of the Cu bridges: sum of the resistances of all i pieces (varying geometries and temperatures). L_i, A_i are the length, cross-sectional area of each Cu piece.
$I_{opt,P}$	Current at maximum power
P_{max}	Maximum power output
η_{max}	Maximum efficiency

Note that all material properties (ρ, S, κ) in Equations (8)–(18) are actually temperature averages, e.g., $\alpha_p = \frac{1}{\Delta T} \int_{T_c}^{T_h} \alpha_p(T) dT$, $\rho_p = \frac{1}{\Delta T} \int_{T_c}^{T_h} \rho_p(T) dT$ and $\kappa_p = \left(\frac{1}{\Delta T} \int_{T_c}^{T_h} \kappa_p^{-1}(T) dT \right)^{-1}$.

When modeling module properties, the resistance of the bridges is often neglected, [22,59] as in our previous work [29]. Here, R_{Cu} was considered explicitly in Equation (14), and represents 4% of the calculated total resistance. This not-so-small value arises despite a relatively large bridge thickness of 250 µm because of its relatively long length L_i and the significantly reduced conductivity of the Cu at the hot side temperature [60]. Note also that we have taken the total bridge length; which, systematically overestimates the effective length. The dimensions of the hot side bridges are L_h = 12 mm and A_h = 6 mm; those of the shorter bridge on the cold side are $L_{c,s}$ = 12 mm and $A_{c,s}$ = 5 mm; and those of the two longer bridges are $L_{c,l}$ = 16 mm and $A_{c,l}$ = 5 mm. r_c = 4 µΩcm^2 is used for all legs, as measured and reported in [29]. Similarly to [29], temperatures at $I_{opt,P}$ are used also for η_{max} calculations because the difference between $I_{opt,P}$ and $I_{opt,\eta}$ < 7%, so temperature conditions at both currents are similar.

3. Results

The Seebeck coefficient, electrical resistivity and thermal conductivity profiles of the n-type legs are obtained from $n(x)$ and $T(x)$. The procedure is represented in Figure 4.

From the calculated profiles, average material properties are calculated for the n-type leg; while for the p-type, a homogeneous material is assumed. The averages are obtained in dependence of the relevant temperature interval ($\Delta T_{TE,I}$), obtained using the CPM with the equations presented above, following, e.g., the procedure outlined in [29]. Module parameters are also calculated following those equations. A comparison of the relevant quantities assuming a homogeneous n-type leg and an inhomogeneous leg using the CPIM is presented in Figure 5.

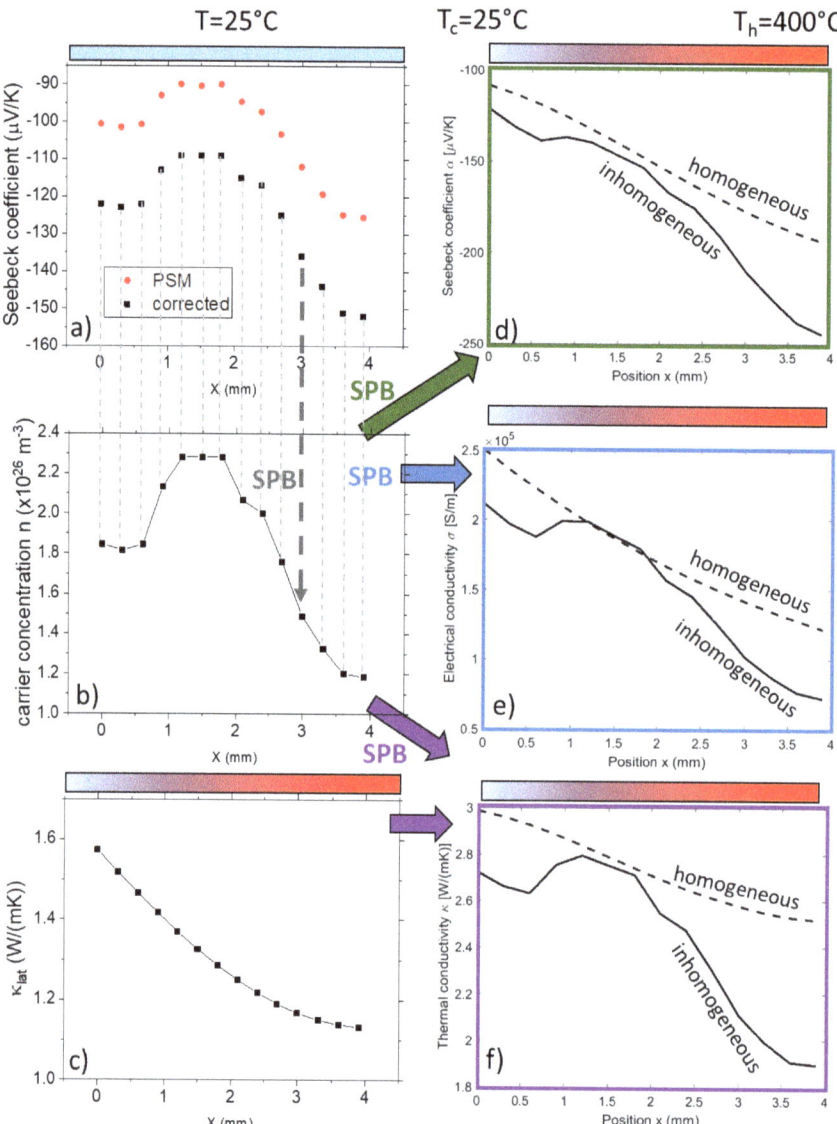

Figure 4. Schematics of the procedure for the calculation of the property profiles for an inhomogeneous leg: (**a**) exemplary line scan of the Seebeck coefficient at room temperature, measured with PSM and also corrected for the cold finger effect; (**b**) carrier concentration (spatial) profile obtained from (**a**) using the SPB model; (**c**) fitted lattice thermal conductivity from experimental data of the sample directly after sintering; (**d**) corresponding Seebeck coefficient profile (in a temperature gradient) calculated using Equation (1); (**e**) corresponding electrical resistivity profile calculated using Equation (6); (**f**) corresponding thermal conductivity profile calculated using Equation (7). A comparative profile for a homogeneous material using the properties directly after sintering is added in dashed lines in (**d–f**). In (**c–f**), a linear temperature profile is assumed between $T_c = 25\ °C$ and $T_h = 400\ °C$ (for illustration purposes) to convert the temperature dependence of the properties into a spatial profile.

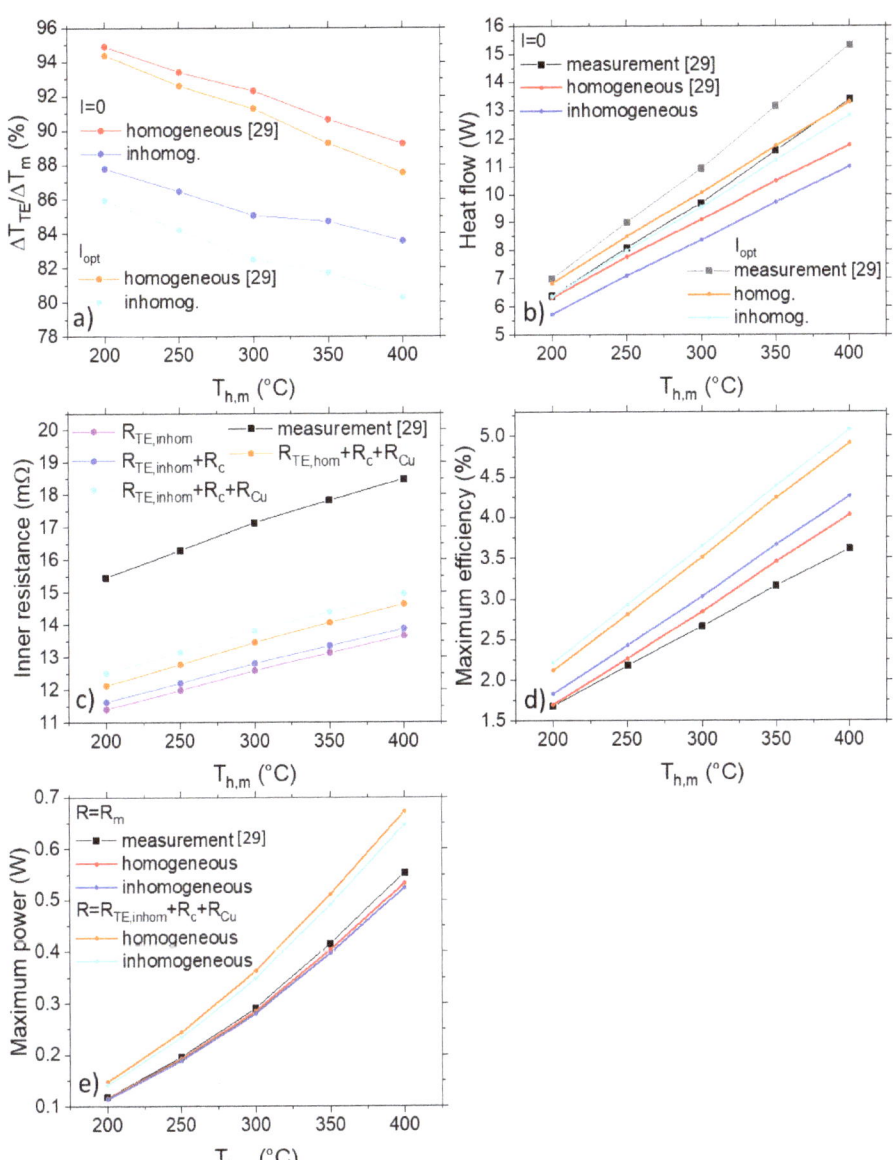

Figure 5. Comparison of experimental data for a Mg$_2$(Si,Sn) TEG (measured during cooling sequence under stabilized temperature conditions, taken from [29]) to CPM and CPIM calculated data for a TEG with homogeneous and inhomogeneous n-type legs, respectively: (**a**) ratio of the temperature difference at the TE legs and the measured temperature difference at the module/heat flow meter interface at $I = 0$ and $I_{opt,P}$; (**b**) heat flows at $I = 0$ and $I_{opt,P}$; (**c**) inner resistance; (**d**) maximum conversion efficiency for measured inner resistance R_m and several calculated inner resistances (considering TE resistance, contact resistance such as $r_c = 4$ µΩcm^2 and R_{Cu}); (**e**) maximum power for measured and calculated inner resistances. The legend in (**e**) also applies in (**d**). $T_{h,m}$ is the temperature at the hot side of the TEG in the measurement column.

Going from the homogeneous material to the inhomogeneous material, carrier loss near the metallization interfaces leads to an increase of the average absolute Seebeck coefficient and electrical resistivity while the thermal conductivity decreases. This trend is observed in the CPM results shown in Figure 5, where the temperature difference at the TE legs (based on measured open-loop voltage of the TEG and the input Seebeck coefficient) and the heat flows decrease for the inhomogeneous legs. Note that the significant decrease in heat flow (~10%) is due to a combined effect of the changed temperatures and the decreased K.

Employing Equation (14), the expected resistance of the TEG can be calculated from material properties, measured contact resistances before device making, and the contribution of the Cu bridges. If we consider the inhomogeneous leg properties the portion of the measured resistance that is due to the TE materials is 74%; while, it would be 70% if we consider the homogeneous properties of the original material. The contributions from contacts and bridge are visible, but relatively small. While the unexplained differences between measured and calculated resistances are large in both cases, this shows that incorrectly assuming homogeneous material would lead to an overestimation of these parasitic resistances. This corresponds to an overestimation of the "effective" contact resistances obtained from $r_{c,\text{eff}} = \frac{(R_m - R_{\text{legs}} - R_{\text{Cu}}) A_n}{2N(1 + A_n/A_p)}$ [29] in the CPM model (at T_h = 200 °C: 65 $\mu\Omega$cm^2 for the homogeneous material, 58 $\mu\Omega$cm^2 for the inhomogeneous material). The relative difference in maximum power between inhomogeneous and homogeneous material (Figure 5e) is smaller for $R = R_m$ than for $R = R_{\text{ideal}} = R_{\text{TE}} + R_c$ with $r_c = 4\ \mu\Omega$cm^2; since for the latter, the resistance value is based on the calculated resistance of the TE materials, which differs; while for the first, the same measured resistance value is used for both materials.

The proportion of parasitic electrical resistance is large in both cases; this confirms that crack formation and propagation are responsible for the suboptimal performance of the TEG, as initially reported in [29]. By avoiding cracking, the maximum power output could be increased by 26%, with a corresponding power density of 1.06 W/cm^2. The smaller relative temperature difference for the inhomogeneous material also indicates that the thermal coupling between the heat source and TE material is worse than initially expected. From an application perspective, thermal losses inside the module and at the heat exchanger need to be minimized [61]. These are not reported for most module characterizations but will be relevant for system optimization and, hence, need to be characterized accurately.

It can be noted that the deviation from measured values for the maximum power in Figure 5e is similar for the homogeneous and for the inhomogeneous legs (2–4% and 3–5%, respectively, with increasing temperature); while, the inhomogeneous leg should be the most realistic case. This is due to the adjustment of the term $(\alpha_p - \alpha_n)\Delta T_{\text{TE,opt}}$ to the measured $U_{0,m}$ value; which, is common to both cases. It is expected that the CPIM error increases for higher temperatures where the SPB model is not as reliable, as the minority carriers start contributing to the conduction [62]. This limitation is even more pronounced for lower carrier concentrations where the conduction regime transition happens at a lower temperature. The hot-side portion of the leg is therefore where the SPB limitations and uncertainty are most relevant due to amplified bipolar effects; since, this portion combines higher temperature and lower carrier concentration. This will be further discussed below.

The calculated heat flow remains below the measured one and the deviation increases from 2–15% to 9–16% when going from a model considering a homogeneous to an inhomogeneous leg, due to the decrease of thermal conductivity with decreasing average carrier concentration of the material. In the case of inhomogeneous material, the deviation between the CPM-predicted heat flow and the measured value can therefore be occasionally larger than the estimated measurement uncertainty (13.5%). As explained in [63], the measurement uncertainty was obtained by testing commercial TEGs which have a larger number of legs, a higher filling factor, and a wider geometry [63]; therefore, it does not necessarily strictly apply to our TEG. As discussed in the SI of [29], the main challenge for small TEG prototypes is the radiative thermal bypass; which, could happen between the hot side and the cold side of the TEG itself, but also between the heater and the soldered cables and the heat flow meter (HFM) where the output heat flow is measured. Even though it cannot be stated with sufficient certainty, the improved analysis also indicates that the results of the efficiency measurements are systematically too low, i.e., the TEG performance is underestimated, as the otherwise quite predictive CPM model disagrees with the experimental data for heat flow and efficiency.

4. Discussion

Figure 6a shows the Seebeck coefficient with respect to temperature for experimental data and calculated data with the SPB model for different carrier concentrations. It can be seen that beyond a certain temperature the experimental data starts to bend much more than the SPB model due to the minority carrier contribution. This bending indicates the maximum temperature at which the SPB is reliable for each carrier concentration. The carrier concentration range shown in Figure 6a corresponds to the range determined in the inhomogeneous material, with lower carrier concentrations at the hot and cold sides and higher carrier concentrations in the middle portion. It can be therefore seen that, on an inhomogeneous leg with a gradient from 25 °C to 400 °C, the SPB prediction of the portion between approximately 350 °C and 400 °C would not be accurate.

Figure 6b,c show the calculated resulting profiles of the n-type material power factor (PF) and figure of merit (zT) assuming a linear temperature profile between 25 °C and 400 °C. As mentioned above, for a fraction of the leg on the hot side, the SPB data is likely not reliable; however, it can be seen that also the rest of the leg, the inhomogeneous leg is predicted to have a higher performance. This is only on a first glance in contradiction to the lower power CPIM calculation of the inhomogeneous leg in Figure 5e, as the reduced ΔT_{TE} for the inhomogeneous material overcompensates the increase in α. Efficiency, on the other hand, is also governed by the heat flow, which is significantly lower in the inhomogeneous leg, leading to the increased efficiency with CPIM shown in Figure 5d.

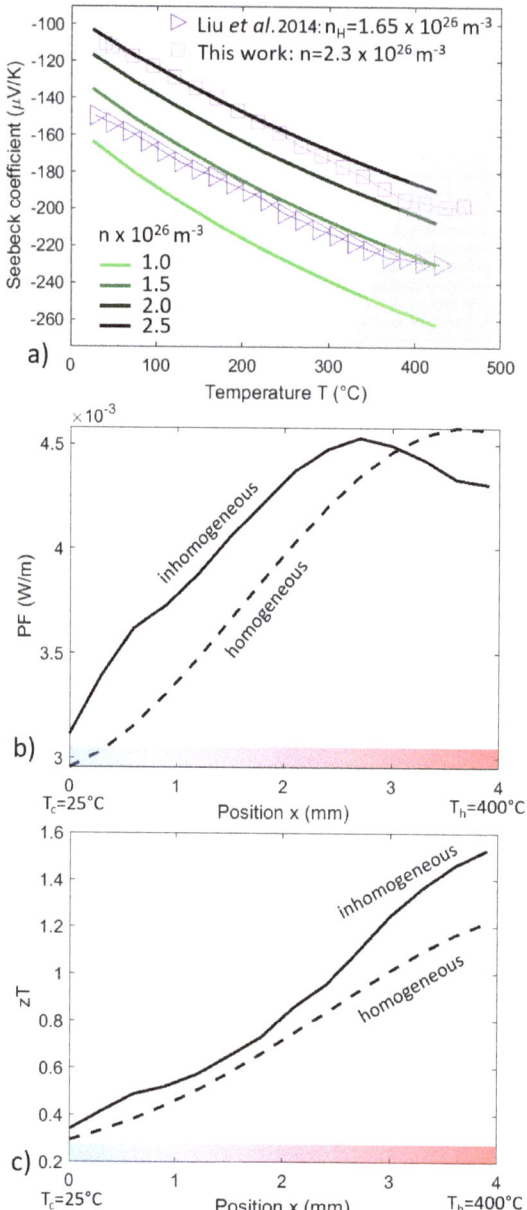

Figure 6. (**a**) Seebeck coefficient with respect to temperature for: experimental data (from this work and the literature [64]) and calculated data for different carrier concentrations, obtained with the SPB model. It was added for visualization although no perfect match is expected as SPB parameters were obtained by different synthesis routes and for different sample compositions. The grey area visualizes the temperature range in which the SPB model apparently deviates from experimental data. Calculation of the (**b**) PF and (**c**) zT profiles for an inhomogeneous (SPB data) and a homogeneous leg (experimental data) with an assumed linear temperature profile between T_c = 25 °C and T_h = 400 °C.

The SPB predictions in Figure 6b,c show that the Zn-induced gradients initially present on the n-type legs after contacting are not detrimental to the TEG's performance. The as-sintered material was not synthesized intentionally with a lower carrier concentration from the start because the SPB model tends to overestimate the figure of merit for low n (see, e.g., [50,65]). In practice, a lower-doped material might therefore have a lower performance. Also, the material was synthesized aiming for a maximized zT_{max}; while zT_{avg} is actually a better performance indicator [58].

In this work, a constant mobility parameter is assumed in the SPB model, independently of the carrier concentration value. If the local change in carrier concentration is due to Mg loss in the TE material (diffusion into the metallization), it could also have an effect on the carrier mobility, as was previously observed in [31,55]. For a one-dimensional current flow along the leg holds $\frac{\partial U}{\partial x} \propto \frac{1}{c}$; an approximately constant mobility can therefore be verified by comparing the relative change in the slope of the electrical potential along a leg on a gradient portion and a middle portion of a leg as shown in Figure 3a in the light and dark grey zones (the slopes are, respectively, 8.3 mV/m and 3.5 mV/m) to the relative change of carrier concentration in those portions ($8.5 \cdot 10^{25}$ m^{-3} and $1.7 \cdot 10^{26}$ m^{-3}, respectively). In our case, both relative changes are about a factor of ≈ 2. This means that the electrical resistivity and the carrier concentration changed quite proportionally; therefore, mobility stayed relatively constant. This analysis validates the use of Equation (5) with a constant E_{Def} for the inhomogeneous material and is of significant importance, as a change in mobility is one of the largest uncertainties of the SPB model.

Besides the limitations of the SPB model, part of the calculation error could also originate from the specific methodology of the CPIM, especially the determination of the corrected Seebeck coefficient profile, see the method section. Finally, the CPM is based on temperature averaging, which balances out the uncertainty on the carrier concentration profile; it is therefore, by definition, not sensitive to fine-tuning of the TE properties.

Lastly, the material change could have an impact on the self-compatibility of the material. This can be verified by calculating the compatibility factor $s(T) = \frac{(-1+\sqrt{1+zT})}{\alpha T}$ [38] for the minimum and maximum carrier concentration values of the profile; which, gives 4.4 for both carrier concentrations at 200 °C and 3.8 and 3.9 at 400 °C, respectively. Limited compatibility becomes an issue when s differs by a factor of 2 [38]; it, therefore, is not expected to play a role here.

As TEGs are supposed to have a long-term application, a method of analysis that considers degradation mechanisms is crucial to understand device behavior. The new method presented here, CPIM, is easy to implement, as SPB-based material models are available for several relevant thermoelectric material systems [54,65,66] and the often-implemented CPM methodology needs to be adjusted only marginally. A further improvement of the presented approach would be to employ a two-band model [67] to overcome the limitations of the SPB model and Finite Element Modelling to cross-check and deepen the understanding of the results presented in this work.

5. Conclusions

In this paper, we present a new coupled material–device modeling approach, the CPIM, which is a methodology developed to capture property inhomogeneity in TE legs and which adapts the CPM model accordingly. This concept can be transferred to cases where the inhomogeneity is by design, for example for graded materials [38]. The CPIM furthermore enables an analysis of TEGs that have seen a thermal load where assuming homogeneous material could lead to wrong conclusions. For the considered case of a measured Mg$_2$(Si,Sn) based TEG, we obtain a smaller temperature difference at the TE legs and a reduced heat flow, compared to the initially assumed homogeneous material, beyond the assumptions and doubts of the SPB model. Consequently, our deepened analysis of the experimental data allows us to identify the heat flow measurement as a main future challenge for accurate measurement of small TEG prototypes.

Author Contributions: Conceptualization, E.M. and J.d.B.; Methodology, J.C.; Formal analysis, J.C.; Investigation, J.C.; Writing—original draft, J.C.; Writing—review & editing, J.C. and J.d.B.; Visualization, J.C.; Supervision, E.M. and J.d.B.; Project administration, E.M. and J.d.B.; Funding acquisition, E.M. and J.d.B. All authors have read and agreed to the published version of the manuscript.

Funding: This research was co-funded by DAAD fellowship program No. 57424731 and the Deutsche Forschungs-gemeinschaft (DFG, German Research Foundation), project number 396709363.

Data Availability Statement: The data presented in this study are available on request from the corresponding author. The data are easily reproduceable by the described methodology.

Acknowledgments: The authors would like to gratefully acknowledge the endorsement for the DLR executive Board Members for Space Research and Technology, as well as the financial support from the Young Research Group Leader Program. We would also like to acknowledge Pawel Ziolkowski for his precious help with the PSM analysis and reviewing the manuscript, Przemyslaw Blaschkewitz for his help with the thermoelectric measurements, as well as Kamila Hasbuna and Harshita Naithani for their help with the SPB MATLAB scripts.

Conflicts of Interest: The authors declare no conflict of interest.

References

1. Twaha, S.; Zhu, J.; Yan, Y.; Li, B. A comprehensive review of thermoelectric technology: Materials, applications, modelling and performance improvement. *Renew. Sustain. Energy Rev.* **2016**, *65*, 698–726. [CrossRef]
2. Karellas, S.; Leontaritis, A.-D.; Panousis, G.; Bellos, E.; Kakaras, E. Energetic and exergetic analysis of waste heat recovery systems in the cement industry. *Energy* **2013**, *58*, 147–156. [CrossRef]
3. Jaziri, N.; Boughamoura, A.; Müller, J.; Mezghani, B.; Tounsi, F.; Ismail, M. A comprehensive review of Thermoelectric Generators: Technologies and common applications. *Energy Rep.* **2020**, *6*, 264–287. [CrossRef]
4. Ebling, D.G.; Krumm, A.; Pfeiffelmann, B.; Gottschald, J.; Bruchmann, J.; Benim, A.C.; Adam, M.; Labs, R.; Herbertz, R.R. Development of a System for Thermoelectric Heat Recovery from Stationary Industrial Processes. *J. Electron. Mater.* **2016**, *45*, 3433–3439. [CrossRef]
5. Li, K.; Garrison, G.; Zhu, Y.; Moore, M.; Liu, C.; Hepper, J.; Bandt, L.; Horne, R.N.; Petty, S. Thermoelectric power generator: Field test at Bottle Rock geothermal power plant. *J. Power Sources* **2021**, *485*, 229266. [CrossRef]
6. Allison, L.K.; Andrew, T. A Wearable All-Fabric Thermoelectric Generator. *Adv. Mater. Technol.* **2019**, *4*, 1800615. [CrossRef]
7. Elmoughni, H.M.; Menon, A.K.; Wolfe, R.M.W.; Yee, S.K. A Textile-Integrated Polymer Thermoelectric Generator for Body Heat Harvesting. *Adv. Mater. Technol.* **2019**, *4*, 1800708. [CrossRef]
8. Hasan, M.N.; Nafea, M.; Nayan, N.; Mohamed Ali, M.S. Thermoelectric Generator: Materials and Applications in Wearable Health Monitoring Sensors and Internet of Things Devices. *Adv. Mater. Technol.* **2021**, *7*, 2101203. [CrossRef]
9. Gorai, P.; Stevanović, V.; Toberer, E. Computationally guided discovery of thermoelectric materials. *Nat. Rev. Mater.* **2017**, *2*, 17053. [CrossRef]
10. Iversen, B.B. Breaking thermoelectric performance limits. *Nat. Mater.* **2021**, *20*, 1309–1310. [CrossRef]
11. Snyder, G.J.; Toberer, E.S. *Complex Thermoelectric Materials, in Materials for Sustainable Energy: A Collection of Peer-Reviewed Research and Review Articles from Nature Publishing Group*; World Scientific: Singapore, 2011; pp. 101–110.
12. Sankhla, A.; Patil, A.; Kamila, H.; Yasseri, M.; Farahi, N.; Mueller, E.; de Boor, J. Mechanical alloying of optimized Mg$_2$ (Si, Sn) solid solutions: Understanding phase evolution and tuning synthesis parameters for thermoelectric applications. *ACS Appl. Energy Mater.* **2018**, *1*, 531–542. [CrossRef]
13. Kamila, H.; Sankhla, A.; Yasseri, M.; Hoang, N.; Farahi, N.; Mueller, E.; de Boor, J. Synthesis of p-type Mg$_2$Si$_1$-xSnx with x = 0–1 and optimization of the synthesis parameters. *Mater. Today Proc.* **2019**, *8*, 546–555. [CrossRef]
14. Trivedi, V.; Battabyal, M.; Balasubramanian, P.; Muralikrishna, G.M.; Jain, P.K.; Gopalan, R. Microstructure and doping effect on the enhancement of the thermoelectric properties of Ni doped Dy filled CoSb3 skutterudites. *Sustain. Energy Fuels* **2018**, *2*, 2687–2697. [CrossRef]
15. Muthiah, S.; Singh, R.; Pathak, B.; Avasthi, P.K.; Kumar, R.; Kumar, A.; Srivastava, A.; Dhar, A. Significant enhancement in thermoelectric performance of nanostructured higher manganese silicides synthesized employing a melt spinning technique. *Nanoscale* **2018**, *10*, 1970–1977. [CrossRef]
16. Moghaddam, A.O.; Shokuhfar, A.; Zhang, Y.; Zhang, T.; Cadavid, D.; Arbiol, J.; Cabot, A. Ge-Doped ZnSb/β-Zn$_4$Sb$_3$ Nanocomposites with High Thermoelectric Performance. *Adv. Mater. Interfaces* **2019**, *6*, 1900467. [CrossRef]
17. Jood, P.; Male, J.P.; Anand, S.; Matsushita, Y.; Takagiwa, Y.; Kanatzidis, M.G.; Snyder, G.J.; Ohta, M. Na Doping in PbTe: Solubility, Band Convergence, Phase Boundary Mapping, and Thermoelectric Properties. *J. Am. Chem. Soc.* **2020**, *142*, 15464–15475. [CrossRef]
18. Song, K.-M.; Shin, D.-K.; Jang, K.-W.; Choi, S.-M.; Lee, S.; Seo, W.-S.; Kim, I.-H. Synthesis and Thermoelectric Properties of Ce 1−z Pr z Fe 4−x Co x Sb 12 Skutterudites. *J. Electron. Mater.* **2017**, *46*, 2634–2639. [CrossRef]

19. Yu, J.; Xing, Y.; Hu, C.; Huang, Z.; Qiu, Q.; Wang, C.; Xia, K.; Wang, Z.; Bai, S.; Zhao, X. Half-heusler thermoelectric module with high conversion efficiency and high power density. *Adv. Energy Mater.* **2020**, *10*, 2000888. [CrossRef]
20. Chu, J.; Huang, J.; Liu, R.; Liao, J.; Xia, X.; Zhang, Q.; Wang, C.; Gu, M.; Bai, S.; Shi, X. Electrode interface optimization advances conversion efficiency and stability of thermoelectric devices. *Nat. Commun.* **2020**, *11*, 1–8. [CrossRef]
21. Jood, P.; Ohta, M.; Yamamoto, A.; Kanatzidis, M.G. Excessively doped PbTe with Ge-induced nanostructures enables high-efficiency thermoelectric modules. *Joule* **2018**, *2*, 1339–1355. [CrossRef]
22. Ying, P.; He, R.; Mao, J.; Zhang, Q.; Reith, H.; Sui, J.; Ren, Z.; Nielsch, K.; Schierning, G. Towards tellurium-free thermoelectric modules for power generation from low-grade heat. *Nat. Commun.* **2021**, *12*, 1121. [CrossRef]
23. Bode, C.; Friedrichs, J.; Somdalen, R.; Koehler, J.; Büchter, K.-D.; Falter, C.; Kling, U.; Ziolkowski, P.; Zabrocki, K.; Mueller, E.; et al. Thermoelectric Energy Recuperation for Aviation—Project Overview and Potentials. *J. Eng. Gas Turbines Power.* **2017**, *139*, 101201. [CrossRef]
24. Farahi, N.; Stiewe, C.; Truong, D.N.; de Boor, J.; Müller, E. High efficiency Mg$_2$(Si,Sn)-based thermoelectric materials: Scale-up synthesis, functional homogeneity, and thermal stability. *RSC Adv.* **2019**, *9*, 23021–23028. [CrossRef]
25. Kim, H.S.; Kikuchi, K.; Itoh, T.; Iida, T.; Taya, M. Design of segmented thermoelectric generator based on cost-effective and light-weight thermoelectric alloys. *Mater. Sci. Eng. B* **2014**, *185*, 45–52. [CrossRef]
26. de Boor, J.; Dasgupta, T.; Saparamadu, U.; Müller, E.; Ren, Z.F. Recent progress in p-type thermoelectric magnesium silicide based solid solutions. *Mater. Today Energy* **2017**, *4*, 105–121. [CrossRef]
27. Gao, P. *Mg$_2$(Si, Sn)-Based Thermoelectric Materials and Devices*; Michigan State University: East Lansing, MI, USA, 2016; p. 128.
28. Goyal, G.K.; Dasgupta, T. Fabrication and testing of Mg$_2$Si$_{1-x}$Sn$_x$ based thermoelectric generator module. *Mater. Sci. Eng. B* **2021**, *272*, 115338. [CrossRef]
29. Camut, J.; Ziolkowski, P.; Ponnusamy, P.; Stiewe, C.; Mueller, E.; de Boor, J. Efficiency measurement and modelling of a high performance Mg$_2$(Si, Sn)-based thermoelectric generator. *Adv. Eng. Mater.* **2022**, *25*, 2200776. [CrossRef]
30. Kato, D.; Iwasaki, K.; Yoshino, M.; Yamada, T.; Nagasaki, T. Control of Mg content and carrier concentration via post annealing under different Mg partial pressures for Sb-doped Mg$_2$Si thermoelectric material. *J. Solid State Chem.* **2018**, *258*, 93–98. [CrossRef]
31. Sankhla, A.; Kamila, H.; Naithani, H.; Mueller, E.; de Boor, J. On the role of Mg content in Mg$_2$(Si, Sn): Assessing its impact on electronic transport and estimating the phase width by in situ characterization and modelling. *Mater. Today Phys.* **2021**, *21*, 100471. [CrossRef]
32. Ayachi, S.; Radhika, D.; Prasanna, P.; Park, S.; Jaywan, C.; SuDong, P.; Byungki, R.; Eckhard, M.; de boor, J. On the Relevance of Point Defects for the Selection of Contacting Electrodes: Ag as an Example for Mg$_2$(Si,Sn)-based Thermoelectric Generators. *Mater. Today Phys.* **2021**, *16*, 100309. [CrossRef]
33. Pham, N.H.; Farahi, N.; Kamila, H.; Sankhla, A.; Ayachi, S.; Müller, E.; de Boor, J. Ni and Ag electrodes for magnesium silicide based thermoelectric generators. *Mater. Today Energy* **2019**, *11*, 97–105. [CrossRef]
34. Shang, H.; Liang, Z.; Xu, C.; Song, S.; Huang, D.; Gu, H.; Mao, J.; Ren, Z.; Ding, F. N-type Mg$_3$Sb$_2$-xBix with improved thermal stability for thermoelectric power generation. *Acta Mater.* **2020**, *201*, 572–579. [CrossRef]
35. Brož, P.; Zelenka, F.; Kohoutek, Z.; Vřešťál, J.; Vykoukal, V.; Buršík, J.; Zemanová, A.; Rogl, G.; Rogl, P. Study of thermal stability of CoSb$_3$ skutterudite by Knudsen effusion mass spectrometry. *Calphad* **2019**, *65*, 1–7. [CrossRef]
36. Lemine, A.S.; Fayyaz, O.; Shakoor, A.; Ahmad, Z.; Bhadra, J.; Al-Thani, N.J. Corrosion Behavior of Thermoelectric P-and N-Type Bismuth Telluride Alloys Developed through Microwave Sintering Process. Available online: https://ssrn.com/abstract=4257563 (accessed on 28 February 2023).
37. El Oualid, S.; Kogut, I.; Benyahia, M.; Geczi, E.; Kruck, U.; Kosior, F.; Masschelein, P.; Candolfi, C.; Dauscher, A.; Koenig, J.D.; et al. High Power Density Thermoelectric Generators with Skutterudites. *Adv. Energy Mater.* **2021**, *11*, 2100580. [CrossRef]
38. Ponnusamy, P.; Naithani, H.; Müller, E.; de Boor, J. Grading studies for efficient thermoelectric devices using combined 1D material and device modeling. *J. Appl. Phys.* **2022**, *132*, 115702. [CrossRef]
39. Rowe, D. *Conversion Efficiency and Figure-Of-Merit, in CRC Handbook of Thermoelectrics*; CRC Press: Boca Raton, FL, USA, 1995; pp. 31–37.
40. Goupil, C. *Continuum Theory and Modeling of Thermoelectric Elements*; John Wiley & Sons: Hoboken, NJ, USA, 2015.
41. Ioffe, A.F.; Stil'Bans, L.S.; Iordanishvili, E.K.; Stavitskaya, T.S.; Gelbtuch, A.; Vineyard, G. Semiconductor thermoelements and thermoelectric cooling. *Phys. Today* **1959**, *12*, 42. [CrossRef]
42. Ponnusamy, P.; de Boor, J.; Müller, E. Using the constant properties model for accurate performance estimation of thermoelectric generator elements. *Appl. Energy* **2020**, *262*, 114587. [CrossRef]
43. Borup, K.A.; De Boor, J.; Wang, H.; Drymiotis, F.; Gascoin, F.; Shi, X.; Chen, L.; Fedorov, M.I.; Müller, E.; Iversen, B.B. Measuring thermoelectric transport properties of materials. *Energy Environ. Sci.* **2015**, *8*, 423–435. [CrossRef]
44. De Boor, J.; Stiewe, C.; Ziolkowski, P.; Dasgupta, T.; Karpinski, G.; Lenz, E.; Edler, F.; Mueller, E. High-temperature measurement of Seebeck coefficient and electrical conductivity. *J. Electron. Mater.* **2013**, *42*, 1711–1718. [CrossRef]
45. Vaney, J.B.; Piarristeguy, A.; Ohorodniichuck, V.; Ferry, O.; Pradel, A.; Alleno, E.; Monnier, J.; Lopes, E.B.; Gonçalves, A.P.; Delaizir, G.; et al. Effective medium theory based modeling of the thermoelectric properties of composites: Comparison between predictions and experiments in the glass–crystal composite system Si$_{10}$As$_{15}$Te$_{75}$–Bi$_{0.4}$Sb$_{1.6}$Te$_3$. *J. Mater. Chem. C* **2015**, *3*, 11090–11098. [CrossRef]

46. Webman, I.; Jortner, J.; Cohen, M.H. Thermoelectric power in inhomogeneous materials. *Phys. Rev. B* **1977**, *16*, 2959–2964. [CrossRef]
47. Hao, Q.; Zhao, H.; Xiao, Y.; Xu, D. Thermal investigation of nanostructured bulk thermoelectric materials with hierarchical structures: An effective medium approach. *J. Appl. Phys.* **2018**, *123*, 014303. [CrossRef]
48. Castillo-Hernández, G.; Müller, E.; de Boor, J. Impact of the Dopant Species on the Thermomechanical Material Properties of Thermoelectric $Mg_2Si_{0.3}Sn_{0.7}$. *Materials* **2022**, *15*, 779. [CrossRef]
49. De Boor, J.; Müller, E. Data analysis for Seebeck coefficient measurements. *Rev. Sci. Instrum.* **2013**, *84*, 065102. [CrossRef]
50. Kamila, H.; Sahu, P.; Sankhla, A.; Yasseri, M.; Pham, H.-N.; Dasgupta, T.; Mueller, E.; de Boor, J. Analyzing transport properties of p-type Mg_2Si–Mg_2Sn solid solutions: Optimization of thermoelectric performance and insight into the electronic band structure. *J. Mater. Chem. A* **2019**, *7*, 1045–1054. [CrossRef]
51. Martin, J. Apparatus for the high temperature measurement of the Seebeck coefficient in thermoelectric materials. *Rev. Sci. Instrum.* **2012**, *83*, 065101. [CrossRef]
52. Ziolkowski, P.; Stiewe, C.; de Boor, J.; Druschke, I.; Zabrocki, K.; Edler, F.; Haupt, S.; König, J.; Mueller, E. Iron Disilicide as High-Temperature Reference Material for Traceable Measurements of Seebeck Coefficient between 300 K and 800 K. *J. Electron. Mater.* **2017**, *46*, 51–63. [CrossRef]
53. Camut, J.; Ayachi, S.; Castillo-Hernández, G.; Park, S.; Ryu, B.; Park, S.; Frank, A.; Stiewe, C.; Müller, E.; de Boor, J. Overcoming Asymmetric Contact Resistances in Al-Contacted $Mg_2(Si, Sn)$ Thermoelectric Legs. *Materials* **2021**, *14*, 6774. [CrossRef]
54. May, A.F.; Snyder, G.J. Introduction to Modeling Thermoelectric Transport at High Temperatures, in Materials, Preparation, and Characterization in Thermoelectrics; CRC Press: Boca Raton, FL, USA, 2017; pp. 207–224.
55. Sankhla, A.; Kamila, H.; Kelm, K.; Mueller, E.; de Boor, J. Analyzing thermoelectric transport in n-type $Mg_2Si_{0.4}Sn_{0.6}$ and correlation with microstructural effects: An insight on the role of Mg. *Acta Mater.* **2020**, *199*, 85–95. [CrossRef]
56. Ziolkowski, P.; Karpinski, G.; Dasgupta, T.; Mueller, E. Probing thermopower on the microscale. *Phys. Status Solidi* **2013**, *210*, 89–105. [CrossRef]
57. Platzek, D.; Karpinski, G.; Stiewe, C.; Ziolkowski, P.; Drasar, C.; Muller, E. Potential-Seebeck-microprobe (PSM): Measuring the spatial resolution of the Seebeck coefficient and the electric potential. In In Proceedings of the ICT 2005, 24th International Conference on Thermoelectrics, Clemson, SC, USA, 19–23 June 2005; IEEE: Piscataway, NJ, USA, 2005.
58. Ponnusamy, P.; Kamila, H.; Müller, E.; de Boor, J. Efficiency as a performance metric for material optimization in thermoelectric generators. *J. Phys. Energy* **2021**, *3*, 044006. [CrossRef]
59. Goyal, G.K.; Dasgupta, T. Generic Approach for Contacting Thermoelectric Solid Solutions: Case Study in n- and p-Type $Mg_2Si_{0.3}Sn_{0.7}$. *ACS Appl. Mater. Interfaces* **2021**, *13*, 20754–20762. [CrossRef]
60. Yang, P.; Dai, S.; Ma, T.; Huang, A.; Jiang, G.; Wang, Y.; Hong, Z.; Jin, Z. Analysis of Peak Electromagnetic Torque Characteristics for Superconducting DC Induction Heaters. *IEEE Access* **2020**, *8*, 14777–14788. [CrossRef]
61. Ziolkowski, P.; Poinas, P.; Leszczynski, J.; Karpinski, G.; Müller, E. Estimation of Thermoelectric Generator Performance by Finite Element Modeling. *J. Electron. Mater.* **2010**, *39*, 1934–1943. [CrossRef]
62. de Boor, J. On the applicability of the single parabolic band model to advanced thermoelectric materials with complex band structures. *J. Mater.* **2021**, *7*, 603–611. [CrossRef]
63. Ziolkowski, P.; Blaschkewitz, P.; Müller, E. Heat flow measurement as a key to standardization of thermoelectric generator module metrology: A comparison of reference and absolute techniques. *Measurement* **2021**, *167*, 108273. [CrossRef]
64. Liu, W.; Chi, H.; Sun, H.; Zhang, Q.; Yin, K.; Tang, X.; Zhang, Q.; Uher, C. Advanced thermoelectrics governed by a single parabolic band: $Mg_2Si_{0.3}Sn_{0.7}$, a canonical example. *Phys. Chem. Chem. Phys.* **2014**, *16*, 6893–6897. [CrossRef]
65. Naithani, H.; Dasgupta, T. Critical Analysis of Single Band Modeling of Thermoelectric Materials. *ACS Appl. Energy Mater.* **2020**, *3*, 2200–2213. [CrossRef]
66. Ravich, Y.I.; Efimova, B.A.; Smirnov, I.A. *Semiconducting Lead Chalcogenides*; Stil'bans, L.S., Ed.; Plenum Press: New York, NY, USA, 1970.
67. Naithani, H.; Muller, E.; de Boor, J. Developing a two-parabolic band model for thermoelectric transport modelling using Mg_2Sn as an example. *J. Phys. Energy* **2022**, *4*, 045002. [CrossRef]

Disclaimer/Publisher's Note: The statements, opinions and data contained in all publications are solely those of the individual author(s) and contributor(s) and not of MDPI and/or the editor(s). MDPI and/or the editor(s) disclaim responsibility for any injury to people or property resulting from any ideas, methods, instructions or products referred to in the content.

Article

Experimental Analysis of the Long-Term Stability of Thermoelectric Generators under Thermal Cycling in Air and Argon Atmosphere

Julian Schwab *, Christopher Fritscher, Michael Filatov, Martin Kober, Frank Rinderknecht and Tjark Siefkes

German Aerospace Center (DLR), Institute of Vehicle Concepts, 70569 Stuttgart, Germany
* Correspondence: julian.schwab@dlr.de

Abstract: It is estimated that 72% of the worldwide primary energy consumption is lost as waste heat. Thermoelectric Generators (TEGs) are a possible solution to convert a part of this energy into electricity and heat for space heating. However, for their deployment a proven long-term operation is required. Therefore, this research investigates the long-term stability of TEGs on system level in air and argon atmosphere under thermal cycling up to 543 K. The layout of the examined test objects resembles a TEG in stack design. The results show that the maximal output power of the test object in air reaches a plateau at 57% of the initial power after 50 cycles caused by an increased electrical resistance of the system. Whereas the test object in argon atmosphere shows no significant degradation of electrical power or resistance. The findings represent a step towards the understanding of the long-term stability of TEGs and can be used as a guideline for design decisions.

Keywords: thermoelectric; generator; module; half-Heusler; long-term stability

Citation: Schwab, J.; Fritscher, C.; Filatov, M.; Kober, M.; Rinderknecht, F.; Siefkes, T. Experimental Analysis of the Long-Term Stability of Thermoelectric Generators under Thermal Cycling in Air and Argon Atmosphere. *Energies* **2023**, *16*, 4145. https://doi.org/10.3390/en16104145

Academic Editor: Diana Enescu

Received: 23 March 2023
Revised: 28 April 2023
Accepted: 29 April 2023
Published: 17 May 2023

Copyright: © 2023 by the authors. Licensee MDPI, Basel, Switzerland. This article is an open access article distributed under the terms and conditions of the Creative Commons Attribution (CC BY) license (https:// creativecommons.org/licenses/by/ 4.0/).

1. Introduction

The improvement of the energy efficiency of various technical processes is an essential objective in the commercial and private sector. It reduces emissions of greenhouse gases, counteracts rising energy costs and enables the adherence to legal requirements. Thermal processes have a high potential for efficiency improvements. They are used for processing heat, home heating systems, stationary and mobile internal combustion engines and other uses. It is estimated that currently 72% of the worldwide primary energy consumption is lost as waste heat [1]. Efforts for waste heat recovery are often abandoned due to the complexity or the investment costs of the available systems.

Thermoelectric Generators (TEGs) are a possible solution to recover a higher share of the waste heat. They work according to the Seebeck effect. It describes two dissimilar conductors which are connected electrically in series and thermally in parallel. When the junctions are maintained at a temperature difference an electrical potential across the conductors develops [2]. To utilize this effect, legs of p- and n-type semiconductors are electrically connected and assembled on electric isolators. The assembled devices are Thermoelectric Modules (TEMs). The thermoelectric efficiency of these semiconducting materials is classified by the figure of merit

$$ZT = \frac{\alpha^2 \sigma}{\lambda} T \qquad (1)$$

with the differential Seebeck-coefficient α in VK^{-1}, the electrical conductivity σ in Sm^{-1}, the thermal conductivity λ in $Wm^{-1}K^{-1}$ and the temperature in K [2]. For low temperature applications below 600 K commonly used materials are based on bismuth telluride (BiTe). For applications with higher temperatures, the materials used include lead telluride (PbTe), Skutterudites or half-Heusler compounds. Other materials in the medium temperature region, such as copper sulfide compounds, are currently under development [3].

TEGs are systems consisting of TEMs between the hot side and cold side heat exchangers as shown in Figure 1. They can be integrated in waste heat flows, where the waste heat is absorbed by the hot side heat exchanger and transferred through the TEMs to a fluid in the cold side heat exchangers. In this way electrical energy is produced and additionally in stationary applications the heat in the fluid can be used for space heating. TEGs do not need any moving parts and are therefore low-maintenance and low-noise, compact, and have potentially low investment costs. They are suitable for a variety of applications, with economic benefits even with low to medium amounts of waste heat. Their potential is shown by Kober et al. [4,5] for passenger vehicles, by Heber et al. [6,7] for commercial vehicles and by Schwab et al. [8] for cogeneration in residential heating systems.

Figure 1. Concept of a stackable TEG design in crossflow arrangement [4]. The heat is transferred from the hot gas flow through the Thermoelectric Modules to the coolant flow with the heat exchangers.

An important criterion for the application of TEGs is their long-term stability. It influences their economic benefits and emission reduction potential. Ageing mechanisms of thermoelectric materials are sublimation, oxidation, diffusion processes, mechanical damaging and others. The processes depend on various characteristics, such as the material class and composition, geometry and external conditions. Parameters that influence the long-term stability are their maximal operational temperature, cycling of the temperature, including amplitude and heating rate of the cycles, mechanical parameters, such as compression pressure, or the atmosphere around the TEMs. They impact the thermoelectric properties and therefore the efficiency. Possible prevention strategies include coating of the materials and electrodes, sealing of the TEMs or the operation in a non-reactive atmosphere like vacuum or argon.

Sublimation is the phase change from solid to gaseous state and often degrades thermoelectric material at high temperatures. Sublimation rates of PbTe and PbSnTe increase with the temperature as measured by Bates and Weinstein in vacuum at temperatures of 673 K and 873 K [9]. Similar behavior is reported by Ohsugi et al. for other telluride-based materials. Telluride is dissociated and agglomerates at the surface of the material where it sublimates at temperatures above 673 K [10]. Skutterudites based on $CoSb_3$ also show thermal degradation. The material in varying atmospheres including vacuum, helium and air at temperatures of 293 K to 1123 K is investigated by Leszczynski et al. [11] and Broz et al. [12]. The $CoSb_3$ is stable at temperatures to around 873 K, above this temperature $CoSb_2$ dissociates to the surface and eventually CoSb dissipates. Half-Heusler compounds are reported to be more stable. Experiments by Zelenka et al. with $TiFe1.33Sb$ and $Ti_xNb_{1-x}FeSb$ show very low sublimation of Sb at temperatures to 873 K in an Argon atmosphere [13].

Oxidation of thermoelectric materials is their reaction with oxygen from surrounding air. For PbTe-based materials, models by Berchenko et al. show that $PbTeO_3$ already starts to form at temperatures of 673 K [14]. The influence of the partial pressure of oxygen on the oxidation of PbTe is analyzed by Chen et al. With a rising partial pressure, the oxidation rate rises as well [15]. Skutterudites are also affected by oxidation. This is reported for $CeFe_4Sb_{12}$ and $YbyCo_4Sb_{12}$ at temperatures over 573 K and 650 K, respectively, by Sklad et al. [16] and Xia et al. [17,18]. To reduce the oxidation of $CoSb_3$, different coatings are investigated by Skomedal et al. and Al_2O_3 is found to be effective with reduced oxidation at 180 thermal cycles to 873 K [19]. Another possibility to reduce the oxidation is by altering the surrounding atmosphere. $La_{0.9}Fe_3CoSb_{12}$ oxidizes on air at temperatures above 673 K, this is prevented by Shin et al. with a vacuum atmosphere and no oxidation is found while tempering the material at 823 K for 100 h [20]. Half-Heusler alloys oxidize at higher temperatures. Different material compositions, such as MNiSn, MCoSb with M = Hf, Zr or Ti and NbSeSb at 873 K for 168 h are investigated by Kang et al. MNiSn and NbSeSb show a higher stability of less than 7% then MCoSb [21]. At even higher temperatures of 1000 K, oxidation of TiNiSn to TiO_2 and Ti_2O_3 [22] and of ZrNiSn to ZrO_2 is found by Appel et al. [23]. Similar behavior is also reported by Zillmann et al. for the commercial materials $Zr_{0.5}Hf_{0.5}CoSb_{0.8}Sn_{0.2}$, $Zr_{0.25}Hf_{0.25}Ti_{0.5}NiSn$ and $Zr_{0.4}Hf_{0.6}NiSn_{0.98}Sb_{0.02}$. It oxidizes in an air atmosphere at 1073 K and is stable in an Argon atmosphere [24].

Diffusion processes occur inside of the materials and especially at the contact interfaces to the electrodes. These processes for PbTe and $Pb_{0.6}Sn_{0.4}Te$ in combination with copper and silver electrodes at temperatures of 673 K for 1000 h and 773 K for 50 h are analyzed by Li et al. It shows that $Pb_{0.6}Sn_{0.4}Te$ reacts with both electrode materials even at low temperatures, but the contact of the PbTe is uniform and with no crack formation. However, copper diffuses in the leg and affects the thermoelectric properties [25]. To reduce this, it is possible to apply a diffusion barrier at the contact interface. A NiTe-layer is used and no diffusion while tempering the material at 823 K for 360 h in vacuum is observed by Ferres et al. [26]. Diffusion barriers are also used for skutterudite materials. CoMo is used by Feng et al. [27] and Nb is used by Chu et al. [28] as a metallization layer on $CoSb_3$ and no significant diffusion during testing is found. A direct bonding method for ZrCoSb- and ZrNiSn-based half-Heusler materials is used by Nozariasbmarz et al. High efficiencies of over 9.5% and a performance reduction of 2.7% after tempering the material at 823 K for 20 h is achieved [29].

The combination of mechanical loads and high temperatures, both often in a cyclic manner, results in mechanical damages such as crack formation, creep or direct mechanical failure. The resulting effects are studied by Al Malki et al. for a ZrNiSn-based half-Heusler alloy. When compressed at a temperature of 873 K, it shows no macroscopic failure or creep, but the beginning of crack formation [30].

In addition to the research on material level, the analysis of the long-term stability on TEM level is also relevant. It expands the scope, including the geometry of the module, implications of the production processes and the interactions of the electric isolators, electrodes and thermoelectric legs. Most studies are conducted regarding BiTe-TEM of different manufacturers, as these are the most commonly available type. Ding et al. [31] find that the power reduces by 37% after 7 h at 483 K, Harish et al. [32] obtain similar results after 300 thermal cycles. Riyadi et al. investigate different thermal cycling heating rates and find that the electrical resistance of the TEM rises with higher heating rates [33]. Merienne et al. also investigate different heating rates but with sealed TEMs. They still find a significant power loss and a correlation to the heating rates [34]. Barako et al. cycle from 253 K to 419 K for 45,000 cycles. The effective figure of merit reduces after 40,000 cycles by 20% and after 45,000 cycles by 97%. The reason is a higher electrical resistance due to crack formation [35]). Salvador et al. investigate PbTe- and $CoSb_3$-based high temperature TEMs sealed with aerogel. The TEMs are exposed to cyclic temperatures from 300 K to 673 K and 723 K and tempered for 200 h at 698 K and 750 K, respectively. The PbTe-based TEMs display a lower long-term stability compared to the $CoSb_3$-based TEMs. They show signs of interlayer diffusion, sublimation and oxidation of the material and the electrical connections [36].

The mentioned research focuses on the long-term stability of thermoelectric materials and modules. However, there is no research on the long-term stability on the system level of TEGs, including TEMs, heat exchangers and structural components, such as compression bolts and plates. This is important for applications because it takes additional effects into account. On the system-level, the mechanical degradation resistance of all components affects the overall performance. Creep or crack formation may result in a lower contact pressure or leaks in the inert gas housing. Furthermore, the loads on the TEMs have not been studied in an application-orientated environment with inhomogeneous temperature gradients or pressure distribution.

Therefore, this research focuses on the long-term stability of TEGs on system-level. The aim is to investigate how the atmosphere influences the long-term stability of such systems in an application-orientated test. This is particularly interesting, as systems with air atmosphere are less complex. Additional factors which are relevant for an application of the systems are represented. The test objects are based on the geometry of TEGs and are able to replicate realistic loads. In line with this, the temperature load is applied cyclically as in most waste heat recovery applications. The evaluation is also system-orientated as the primarily considered parameter is the power output of the system and its long-term progress. The first tests are designed to determine the possibility to systematically improve the power output stability with an argon inert gas atmosphere. The findings can be used for the design process of future TEGs and develop them further towards deployment.

2. Methodology

The layout of the test objects is shown in Figure 2, it is based on a stackable TEG design. The heat is generated by electrical heating cartridges which are embedded in a plate. The cartridges have a maximum operation temperature of 1023 K and a total heat power of 10 kW. On one side of the heating plate is an isolating layer and on the other side a steel sheet with grooves on the surface. For the experiment type K thermocouples are placed in the grooves to measure the surface temperature of the TEMs. Six TEMs are electrically connected in series and placed on the steel sheet. The TEMs are based on (Ti,Zr)NiSn for the n-type and FeNbSb for the p-type and produced by the professional manufacturer Yamaha Corporation. The material is chosen as it has a high maximal temperature, a ZT value of around unity and a high mechanical stability. Each TEM has a nominal power output of 46.1 W^{el} at 450 K temperature difference and an active area of around 1453 mm^2. The TEMs are connected with soldered copper cables. Above the TEMs is a coolant heat exchanger as heat sink. It also has grooves on the surface for thermocouples to measure the cold side temperature of the TEMs. The whole stack is compressed to around two MPa pressure on the TEMs with a structure consisting of two stiff compression plates and four compression bolts. The applied pressure and graphite sheets between each layer are needed to minimize the thermal contact resistance between the components.

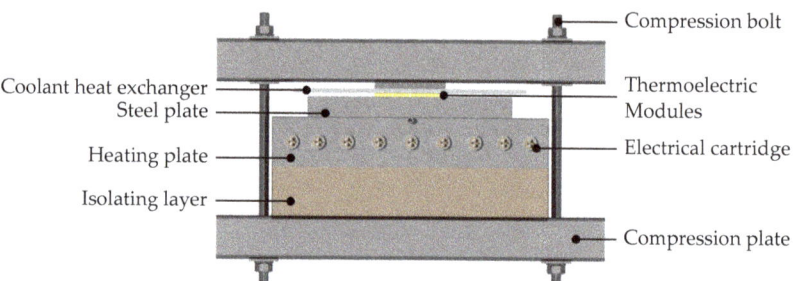

Figure 2. Schematic view of the layout of the test objects. The Thermoelectric Modules are compressed with a compression structure to a heating plate and a coolant heat exchanger.

The experimental setting consists of two identical test objects, which are only distinguishable by their production tolerance. One heating plate is produced with a more accurate tolerance and is thus able to absorb the heat of the cartridges slightly more efficiently. The test objects are placed in gas tight chambers, as shown in Figure 3. One chamber is filled with argon at around 1.5 bar absolute pressure, the other one is open and the test object is exposed to ambient air. The argon concentration of the argon filled chamber is monitored with a lambda probe and its pressure is monitored with an absolute pressure sensor. On the coolant side, the test objects are connected to a temperature control unit with a heat sink. The control unit is able to set the required mass flow and inlet temperature of the water that is used as coolant. The temperature is measured with Pt100 senors in a four wire arrangement in the cooling unit and additionally type K thermocouples directly at the inlet and outlet of the test chambers. Moreover, thermocouples are placed in various places throughout the experimental setup to measure if the system is overheating and to switch off the process. This is particularly important for an automatic procedure.

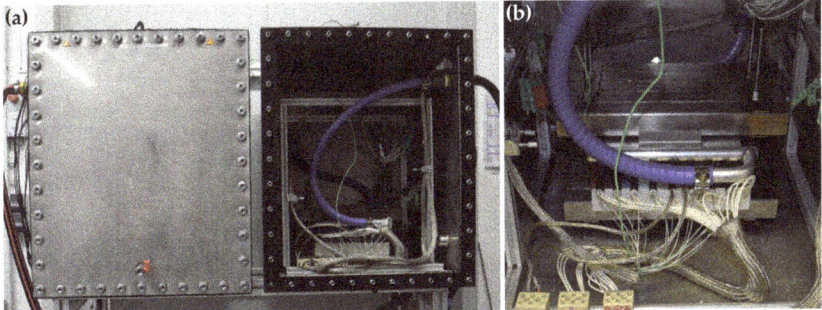

Figure 3. View of the experimental setup: (**a**) Test bench with the closed test chamber filled with argon and the open test chamber. (**b**) Test object in the test chamber.

The TEMs of each test object are connected to a self-developed maximum power point tracker (MPPT). It adjusts the outer electrical resistance of the electrical load to the inner electrical resistance of the TEMs in order to maximize their power output. The tracking algorithm increases and decreases the electrical current on a fixed frequency and adjusts the current accordingly. Other algorithms, such as shuffled frog leaping algorithm [37] or selective harmonic elimination [38], are currently not used. The MPPT has integrated precision resistances for the measurement of voltage and current. The measured values are transferred to the measurement system and are recorded there. The measurement system consists of ProfiMessage data acquisition modules from Delphin Technology. All measured values are supplied by the sensors as an electrical signal. The signal is transferred from the sensor to the measuring device, where it is converted to a digital signal and recorded. This measurement chain is calibrated with a Beamex MC2 device. The output power of the test objects is calculated by multiplying the output voltage and current at the maximum power point. The electrical resistance is calculated by Ohm's law.

For thermal cycling, the cartridges of the first test object start to heat with maximum power until their temperature limit is reached. When their maximum temperature is reached, they are controlled with a pulse-width modulation to keep the temperature constant and to allow a stationary thermal state. During heating, the coolant flow is set to 25 L/min at a temperature of 333 K. When the stationary state is reached, the cartridges stop to heat and the coolant flow is set to 40 L/min at 293 K until the mean cold side temperature is 323 K. Afterwards the heating of the second test object starts in the same pattern. The resulting temperatures are shown in Figure 4. Because of the different production tolerances of the heating plates, the temperature curve and maximum temperature of the test object in air is slightly higher compared to the test object in argon. For the same reason, the heating duration of the test object in air is 34 min and for the one in argon is 43 min. The duration of one complete cycle is around 95 min.

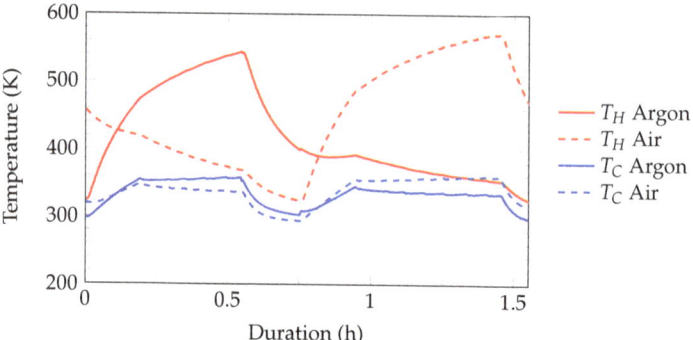

Figure 4. Temperature profiles of the hot and cold side of the test objects in air and argon. One test object is heated while the other test object is cooled and vice versa.

3. Results and Discussion

The thermal cycles are performed until a plateau of the maximum output power is reached. For each cycle, the electric parameters at the maximum power point is measured at a hot side temperature of 543 K. This is the highest hot side temperature both test objects reach during the tests. The results of these measurements are shown in Figure 5. They normalized on the initial values to make them comparable. The initial maximum power of the test object in air is 57.2 W and of the test object in argon is 68.5 W. The initial offset is caused by the differences in production and assembly of the test objects and its components. The pressure and oxygen level of the argon atmosphere is monitored during the experiment and argon is refilled to the initial values when the oxygen level exceeds 5%. The initial values are 1.25 bar and 0.8%, respectively.

The maximum power of the test object in air atmosphere steadily decreases until it reaches a plateau at around 57% of its initial maximum power after around 50 cycles. The course of the measured values of the test object in argon atmosphere shows no steady degradation. After the cycles two and nine, there are abrupt drops in performance, and after cycle 16, there is a shortcut in the system and the power drops significantly. This is caused by a contact of the thermal interface material with the electrodes. At cycle 22, the shortcut is manually removed and the test object returns to operation at around 95% of its initial maximum power with no additional degradation during the subsequent cycles.

Figure 5 shows that the power performance output reduction of the test object in air is caused by an increasing inner resistance from 5.13 Ω to 9.24 Ω and a corresponding decrease in electrical current from 3.38 A to 1.95 A while the voltage is roughly constant at about 16.9 V. Similar to the progression of the power, the electrical values of the test object in argon show no degradation and only the changes due to the shortcut in the system are observable. The inner resistance rises by 4.38% from 4.34 Ω to 4.53 Ω, the voltage rises by 0.12% from 16.95 V to 16.97 V and the current declines by 4.95% from 4.04 A to 3.84 A.

The results show that an air atmosphere causes significantly reduced long-term stability of TEGs compared to the use of an argon atmosphere as inert gas. Although the Seebeck coefficient is not influenced by the air atmosphere for over 50 cycles, the increase of inner resistance leads to a power decrease. Based on the literature on long-term stability of Thermoelectric Modules, important degradation mechanisms are oxidation and crack formation, or a combination of both, on the material and especially at the interfaces to the electrodes; these effects are also found in literature [24]. In Argon gas, there is no oxygen that reacts with the material and causes oxidation; therefore, this effect is limited. Furthermore, the combination of cracking and oxidation in the cracks is limited as well. The mechanical components of the system consisting of the compression bolts, compression plate, and hot and cold side heat exchangers show no degradation during the experiment and there is no positive or negative influence of the overall mechanical system on the long-term stability detected during the measurements.

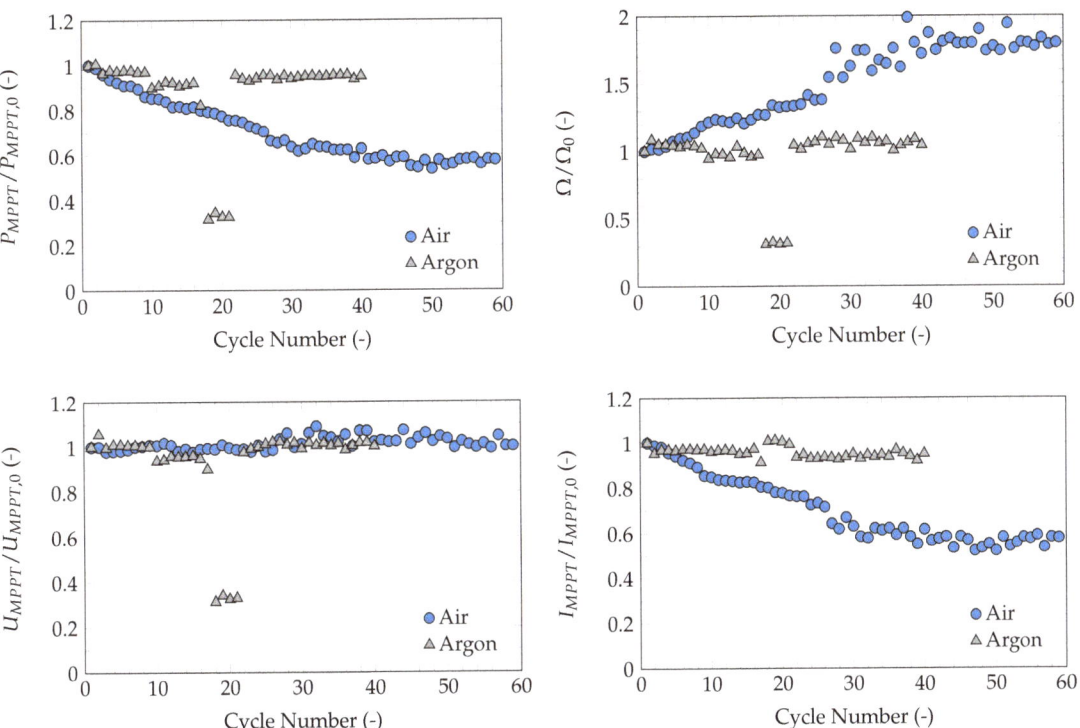

Figure 5. Progression of the electrical parameters power, inner resistance, voltage and current of the test obejts in air and argon. All values are at the maximum power point and normalized to the inital value.

The experiment is designed to represent a real application of TEGs. However, some simplifications are needed for an efficient experimental procedure, that differ to the final applications. These simplifications include a non-symmetry of the stack, as the TEMs are only on one side of the heating plate, the absence of exhaust gas in the hot gas heat exchanger, which may affect its long-term behavior or the absence of the housing, which may also be subject to degradation. Moreover, although the results are already significant, the test procedure should be repeated with the same parameters in order to specify statistic influences, which are based on production tolerances, for example.

The experiment shows the need of an inert gas atmosphere for the successful operation of TEGs with commercially produced high-temperature Thermoelectric Modules. To further understand the systems, experiments with a broader variety of Thermoelectric Modules from different manufacturers or research institutions are needed. Furthermore, future experiments are planned to perform more cycles and to study the influence of parameters such as maximal temperature, cycle duration and oxygen concentration. After future experiments with a higher number of cycles, the Thermoelectric Modules and the overall system should be examined, especially regarding microcrack formation. This research represents a step towards the understanding of the long-term stability of TEGs and can be used as a guideline for design decisions.

4. Conclusions

TEGs are a possible solution to recover waste heat from various processes and to reduce the process costs and emissions. This research aims to investigate their degradation on system level and a possible approach to minimize the degradation. An experiment on two test objects with high-temperature, professionally produced Thermoelectric Modules in air and in argon atmosphere is performed. Their layout is based on a stackable TEG design and resembles its technological features. The hot-side temperature of the test objects is increased to around 543 K and then decreased to 332 K in a cyclic manner. The output power at the maximum power point and the electrical parameters of the test objects are measured.

The results show a significant reduction of the maximal output power of the test object in air reaching a plateau at around 57% of its initial maximum power after 50 cycles. The argon atmosphere shows a positive influence on the degradation behavior, as the output power remains constant during the cyclic tests. In both test objects the voltage is at a constant level, but in air, the inner resistance increases and the current decreases accordingly. Therefore, the increase of the inner resistance leads to the decrease of the output performance. A comparison with the literature shows that oxidation and crack formation on the material and especially at the interfaces to the electrodes are the main reasons for the degradation.

In future experiments, different Thermoelectric Modules from various producers and the influence of test parameters such as cycle duration, maximal temperature and oxygen concentration should be considered. The selection of the different Modules should be based on their maximal temperature, as the long-term stability in the high temperature regime is especially critical. Promising material classes for this investigation are Half-Heusler alloys and Skutterudites. After the experiments, all components should be examined regarding microcrack formation. To consolidate research about this topic, standard testing procedures may be introduced that are based on application-orientated thermal cycles. This research represents a step towards the understanding of the long-term stability of TEGs on system level and can be used as a guideline for design decisions.

Author Contributions: Conceptualization, J.S., C.F. and M.F.; methodology, J.S., C.F. and M.F.; software, J.S., C.F. and M.F.; validation, J.S., C.F. and M.F.; formal analysis, C.F. and M.F.; investigation, J.S., C.F. and M.F.; resources, J.S., C.F., M.F.; data curation, J.S., C.F. and M.F.; writing—original draft preparation, J.S.; writing—review and editing, J.S., C.F., M.F., M.K., F.R. and T.S.; visualization, J.S.; supervision, M.K., F.R. and T.S.; project administration, J.S.; funding acquisition, J.S., M.K., F.R. and T.S. All authors have read and agreed to the published version of the manuscript.

Funding: This research was externally funded by the Federal Ministry for Economic Affairs and Climate Action in the project RecoveryPlus, grant number 03EN2024A.

Data Availability Statement: The data presented in this study are available on request from the corresponding author. The data are not publicly available due to privacy reasons.

Acknowledgments: The authors acknowledge the support and cooperation with all external and internal partners and appreciate their contributions to this work. Especially Takahiro Hayashi from Yamaha Corporation for the helpful discussions and support regarding the thermoelectric modules.

Conflicts of Interest: The authors declare no conflict of interest.

Abbreviations

The following abbreviations are used in this manuscript:

BiTe	Bismuth telluride
C	Cold side
H	Hot side
MPPT	Maximum power point tracker
PbTe	Lead telluride
TEG	Thermoelectric Generator
TEM	Thermoelectric Module

References

1. Forman, C.; Muritala, I.K.; Pardemann, R.; Meyer, B. Estimating the global waste heat potential. *Renew. Sustain. Energy Rev.* **2016**, *57*, 1568–1579. [CrossRef]
2. Rowe, D.M. (Ed.) *Thermoelectrics Handbook: Macro to Nano*; CRC/Taylor & Francis: Boca Raton, FL, USA, 2006.
3. Chen, X.Q.; Fan, S.J.; Han, C.; Wu, T.; Wang, L.J.; Jiang, W.; Dai, W.; Yang, J.P. Multiscale architectures boosting thermoelectric performance of copper sulfide compound. *Rare Met.* **2021**, *40*, 2017–2025. [CrossRef] [PubMed]
4. Kober, M. Holistic Development of Thermoelectric Generators for Automotive Applications. *J. Electron. Mater.* **2020**, *49*, 2910–2919. [CrossRef]
5. Kober, M.; Knobelspies, T.; Rossello, A.; Heber, L. Thermoelectric Generators for Automotive Applications: Holistic Optimization and Validation by a Functional Prototype. *J. Electron. Mater.* **2020**, *49*, 2902–2909. [CrossRef]
6. Heber, L.; Schwab, J. Modelling of a thermoelectric generator for heavy-duty natural gas vehicles: Techno-economic approach and experimental investigation. *Appl. Therm. Eng.* **2020**, *174*, 115156. [CrossRef]
7. Heber, L.; Schwab, J.; Knobelspies, T. 3 kW Thermoelectric Generator for Natural Gas-Powered Heavy-Duty Vehicles—Holistic Development, Optimization and Validation. *Energies* **2022**, *15*, 15. [CrossRef]
8. Schwab, J.; Bernecker, M.; Fischer, S.; Seyed Sadjjadi, B.; Kober, M.; Rinderknecht, F.; Siefkes, T. Exergy Analysis of the Prevailing Residential Heating System and Derivation of Future CO_2-Reduction Potential. *Energies* **2022**, *15*, 3502. [CrossRef]
9. Bates, H.E.; Weinstein, M. Sublimation rates In vacuo of PbTe and $Pb_{0.5}Sn_{0.5}Te$ thermoelements. *Adv. Energy Convers.* **1966**, *6*, 177–180. [CrossRef]
10. Ohsugi, I.J.; Tokunaga, D.; Kato, M.; Yoneda, S.; Isoda, Y. Dissociation and sublimation of tellurium from the thermoelectric tellurides. *Mater. Res. Innov.* **2015**, *19*, S5-301–S5-303. [CrossRef]
11. Leszczynski, J.; Wojciechowski, K.T.; Malecki, A.L. Studies on thermal decomposition and oxidation of $CoSb_3$. *J. Therm. Anal. Calorim.* **2011**, *105*, 211–222. [CrossRef]
12. Brož, P.; Zelenka, F.; Kohoutek, Z.; Vřešťál, J.; Vykoukal, V.; Buršík, J.; Zemanová, A.; Rogl, G.; Rogl, P. Study of thermal stability of $CoSb_3$ skutterudite by Knudsen effusion mass spectrometry. *Calphad* **2019**, *65*, 1–7. [CrossRef]
13. Zelenka, F.; Brož, P.; Vřešťál, J.; Buršík, J.; Zemanová, A.; Rogl, G.; Rogl, P. Study of thermal stability of half-Heusler alloys $TiFe1.33Sb$ and $TixNb1-xFeSb$ (x = 0, 0.15) by differential thermal analysis and Knudsen effusion method. *Calphad* **2021**, *74*, 102292. [CrossRef]
14. Berchenko, N.; Fadeev, S.; Savchyn, V.; Kurbanov, K.; Trzyna, M.; Cebulski, J. Pb–Te–O phase equilibrium diagram and the lead telluride thermal oxidation. *Thermochim. Acta* **2014**, *579*, 64–69. [CrossRef]
15. Chen, L.; Goto, T.; Tu, R.; Hirai, T. High-temperature Oxidation Behavior Of PbTe And Oxidation-resistive Glass Coating. In Proceedings of the 16th International Conference on Thermoelectrics (1997), Dresden, Germany, 26–29 August 1997.
16. Sklad, A.C.; Gaultois, M.W.; Grosvenor, A.P. Examination of $CeFe4Sb12$ upon exposure to air: Is this material appropriate for use in terrestrial, high-temperature thermoelectric devices? *J. Alloys Compd.* **2010**, *505*, L6–L9. [CrossRef]
17. Xia, X.; Qiu, P.; Shi, X.; Li, X.; Huang, X.; Chen, L. High-Temperature Oxidation Behavior of Filled Skutterudites $Yb_y Co4Sb12$. *J. Electron. Mater.* **2012**, *41*, 2225–2231. [CrossRef]
18. Xia, X.; Qiu, P.; Huang, X.; Wan, S.; Qiu, Y.; Li, X.; Chen, L. Oxidation Behavior of Filled Skutterudite $CeFe4Sb12$ in Air. *J. Electron. Mater.* **2014**, *43*, 1639–1644. [CrossRef]
19. Skomedal, G.; Kristiansen, N.R.; Engvoll, M.; Middleton, H. Methods for Enhancing the Thermal Durability of High-Temperature Thermoelectric Materials. *J. Electron. Mater.* **2014**, *43*, 1946–1951. [CrossRef]
20. Shin, D.K.; Kim, I.H.; Park, K.H.; Lee, S.; Seo, W.S. Thermal Stability of $La0.9Fe3CoSb12$ Skutterudite. *J. Electron. Mater.* **2015**, *44*, 1858–1863. [CrossRef]
21. Kang, H.B.; Saparamadu, U.; Nozariasbmarz, A.; Li, W.; Zhu, H.; Poudel, B.; Priya, S. Understanding Oxidation Resistance of Half-Heusler Alloys for in-Air High Temperature Sustainable Thermoelectric Generators. *ACS Appl. Mater. Interfaces* **2020**, *12*, 36706–36714. [CrossRef]
22. Appel, O.; Cohen, S.; Beeri, O.; Shamir, N.; Gelbstein, Y.; Zalkind, S. Surface Oxidation of TiNiSn (Half-Heusler) Alloy by Oxygen and Water Vapor. *Materials* **2018**, *11*, 2296. [CrossRef]
23. Appel, O.; Breuer, G.; Cohen, S.; Beeri, O.; Kyratsi, T.; Gelbstein, Y.; Zalkind, S. The Initial Stage in Oxidation of ZrNiSn (Half Heusler) Alloy by Oxygen. *Materials* **2019**, *12*, 1509. [CrossRef]
24. Zillmann, D.; Waag, A.; Peiner, E.; Feyand, M.H.; Wolyniec, A. Thermoelectric and Structural Properties of Zr-/Hf-Based Half-Heusler Compounds Produced at a Large Scale. *J. Electron. Mater.* **2018**, *47*, 1546–1554. [CrossRef]
25. Li, C.C.; Drymiotis, F.; Liao, L.L.; Hung, H.T.; Ke, J.H.; Liu, C.K.; Kao, C.R.; Snyder, G.J. Interfacial reactions between PbTe-based thermoelectric materials and Cu and Ag bonding materials. *J. Mater. Chem. C* **2015**, *3*, 10590–10596. [CrossRef]
26. Ferreres, X.R.; Aminorroaya Yamini, S.; Nancarrow, M.; Zhang, C. One-step bonding of Ni electrode to n-type PbTe—A step towards fabrication of thermoelectric generators. *Mater. Des.* **2016**, *107*, 90–97. [CrossRef]
27. Feng, H.; Zhang, L.; Zhang, J.; Gou, W.; Zhong, S.; Zhang, G.; Geng, H.; Feng, J. Metallization and Diffusion Bonding of $CoSb_3$-Based Thermoelectric Materials. *Materials* **2020**, *13*, 1130. [CrossRef]
28. Chu, J.; Huang, J.; Liu, R.; Liao, J.; Xia, X.; Zhang, Q.; Wang, C.; Gu, M.; Bai, S.; Shi, X.; et al. Electrode interface optimization advances conversion efficiency and stability of thermoelectric devices. *Nat. Commun.* **2020**, *11*, 2723. [CrossRef]

29. Nozariasbmarz, A.; Saparamadu, U.; Li, W.; Kang, H.B.; Dettor, C.; Zhu, H.; Poudel, B.; Priya, S. High-performance half-Heusler thermoelectric devices through direct bonding technique. *J. Power Sources* **2021**, *493*, 229695. [CrossRef]
30. Al Malki, M.M.; Qiu, Q.; Zhu, T.; Snyder, G.J.; Dunand, D.C. Creep behavior and postcreep thermoelectric performance of the n-type half-Heusler alloy Hf0.3Zr0.7NiSn0.98Sb0.02. *Mater. Today Phys.* **2019**, *9*, 100134. [CrossRef]
31. Ding, L.C.; Akbarzadeh, A.; Date, A. Performance and reliability of commercially available thermoelectric cells for power generation. *Appl. Therm. Eng.* **2016**, *102*, 548–556. [CrossRef]
32. Harish, S.; Sivaprahasam, D.; Jayachandran, B.; Gopalan, R.; Sundararajan, G. Performance of bismuth telluride modules under thermal cycling in an automotive exhaust thermoelectric generator. *Energy Convers. Manag.* **2021**, *232*, 113900. [CrossRef]
33. Riyadi, T.W.B.; Utomo, B.R.; Effendy, M.; Wijayanta, A.T.; Al-Kayiem, H.H. Effect of thermal cycling with various heating rates on the performance of thermoelectric modules. *Int. J. Therm. Sci.* **2022**, *178*, 107601. [CrossRef]
34. Merienne, R.; Lynn, J.; McSweeney, E.; O'Shaughnessy, S.M. Thermal cycling of thermoelectric generators: The effect of heating rate. *Appl. Energy* **2019**, *237*, 671–681. [CrossRef]
35. Barako, M.T.; Park, W.; Marconnet, A.M.; Asheghi, M.; Goodson, K.E. Thermal Cycling, Mechanical Degradation, and the Effective Figure of Merit of a Thermoelectric Module. *J. Electron. Mater.* **2013**, *42*, 372–381. [CrossRef]
36. Salvador, J.R.; Cho, J.Y.; Ye, Z.; Moczygemba, J.E.; Thompson, A.J.; Sharp, J.W.; König, J.D.; Maloney, R.; Thompson, T.; Sakamoto, J.; et al. Thermal to Electrical Energy Conversion of Skutterudite-Based Thermoelectric Modules. *J. Electron. Mater.* **2013**, *42*, 1389–1399. [CrossRef]
37. Siadatan, A.; fakhari, M.; Taheri, B.; Sedaghat, M. New fundamental modulation technique with SHE using shuffled frog leaping algorithm for multilevel inverters. *Evol. Syst.* **2020**, *11*, 541–557. [CrossRef]
38. Torkaman, H.; Fakhari, M.; Karimi, H.; Taheri, B. New Frequency Modulation Strategy with SHE for H-bridge Multilevel Inverters. In Proceedings of the 2018 4th International Conference on Electrical Energy Systems (ICEES), Chennai, India, 7–9 February 2018; pp. 157–161. [CrossRef]

Disclaimer/Publisher's Note: The statements, opinions and data contained in all publications are solely those of the individual author(s) and contributor(s) and not of MDPI and/or the editor(s). MDPI and/or the editor(s) disclaim responsibility for any injury to people or property resulting from any ideas, methods, instructions or products referred to in the content.

Article

Thermoelectric Performance Evaluation and Optimization in a Concentric Annular Thermoelectric Generator under Different Cooling Methods

Wenlong Yang [1], Wenchao Zhu [1,2], Yang Yang [1,2,*], Liang Huang [1], Ying Shi [1] and Changjun Xie [1,2,*]

[1] School of Automation, Wuhan University of Technology, Wuhan 430070, China; lahei@whut.edu.cn (W.Y.); zhuwenchao@whut.edu.cn (W.Z.); huangliang@cddiot.com (L.H.); a_laly@163.com (Y.S.)
[2] Hubei Key Laboratory of Advanced Technology for Automotive Components, Wuhan University of Technology, Wuhan 430070, China
* Correspondence: whutyangyang@whut.edu.cn (Y.Y.); jackxie@whut.edu.cn (C.X.)

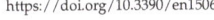

Citation: Yang, W.; Zhu, W.; Yang, Y.; Huang, L.; Shi, Y.; Xie, C. Thermoelectric Performance Evaluation and Optimization in a Concentric Annular Thermoelectric Generator under Different Cooling Methods. *Energies* **2022**, *15*, 2231. https://doi.org/10.3390/en15062231

Academic Editor: Diana Enescu

Received: 19 February 2022
Accepted: 16 March 2022
Published: 18 March 2022

Publisher's Note: MDPI stays neutral with regard to jurisdictional claims in published maps and institutional affiliations.

Copyright: © 2022 by the authors. Licensee MDPI, Basel, Switzerland. This article is an open access article distributed under the terms and conditions of the Creative Commons Attribution (CC BY) license (https:// creativecommons.org/licenses/by/ 4.0/).

Abstract: To ensure effective heat recovery of thermoelectric generators, a cooling system is necessary to maintain the working temperature difference of the thermoelectric couples, which decreases continuously due to thermal diffusion. In order to evaluate and improve the thermoelectric performance of a concentric annular thermoelectric generator under various cooling methods, a comprehensive numerical model of the thermo-fluid-electric multi-physics field for an annular thermoelectric generator with a concentric annular heat exchanger was developed using the finite-element method. The effects of four cooling methods and different exhaust parameters on the thermoelectric performance were investigated. The results show that, in comparison to the cocurrent cooling pattern, the countercurrent cooling pattern effectively reduces temperature distribution non-uniformity and hence increases the maximum output power; however, it requires more thermoelectric semiconductor materials. Furthermore, when using the cocurrent air-cooling method, high exhaust temperatures may result in lower output power; high exhaust mass flow rates result in high exhaust resistance and reduce system net power. The maximum net power output P_{net} = 432.42 W was obtained using the countercurrent water-cooling, corresponding to an optimal thermoelectric semiconductor volume of 9.06×10^{-4} m^3; when compared to cocurrent water-cooling, the maximum net power increased by 8.9%, but the optimal thermoelectric semiconductor volume increased by 21.4%.

Keywords: thermal management; thermoelectric generator; cooling method; annular thermoelectric semiconductor

1. Introduction

The use of waste heat to generate electricity power via thermoelectric generators has been a focus of attention in the field of energy recycling for many years. The majority of the energy produced by fossil fuels is wasted as heat in internal combustion engine vehicles, with only about 30% of the energy converted to usable work. Thermoelectric generators (TEGs) are believed to have the potential and possibility of being used in an automobile's thermal energy recovery system due to their unique advantages, such as no moving parts, no pollution, and the ability to immediately convert thermal energy into electric energy [1,2]. TEG can not only reduce pollution, but also improve fuel efficiency and save energy. The primary components of a thermoelectric generator are a thermoelectric module (TEM), a heat exchanger, and a cooling system. A conventional TEM is a thermoelectric device composed of many rectangular shaped p- and n-type thermoelectric semiconductors electrically connected in series via copper sheets and covered with two ceramic plates. The large amount of high-temperature exhaust gas generated in the exhaust should be cooled in time to ensure normal operation. In addition to the field of automobile exhaust gas recovery, TEGs can also be used effectively in fuel cell systems to recover waste heat and improve overall energy conversion efficiency. Musharavati et al. proposed an integrated system that combines a proton exchange membrane fuel cell with a solar pond system and uses TEG to recover thermal energy, thereby addressing the issues of low thermal efficiency and energy output [3]. Subsequently, they proposed a tandem energy recovery system for proton exchange membrane fuel cells, which uses an organic Rankine cycle and thermoelectric power generation technologies for more energy recovery [4].

However, the current TEGs for waste heat recovery system still suffer from low conversion efficiency, which hinders its commercialization. To improve the performance of TEG, the researchers opened up two main research directions. The first is to enhance the thermoelectric efficiency of thermoelectric semiconductor materials, and the second is to optimize the internal structure of TEG.

The dimensionless constant (ZT) is a common metric for evaluating the efficiency and performance of thermoelectric materials. In order to improve the ZT value of thermoelectric semiconductor materials, significant progress has been made by employing modern synthesis and characterization techniques [5–7], but most thermoelectric materials now still have ZT values in the range of 1–1.6. Bell pointed out that if ZT values of 2 or greater could be achieved, thermoelectric electronic components would be more widely used [8]. Yin et al. achieved a high ZT of 2.2 at 450 °C by alloying $CuBiSe_2$ into GeTe [9]. Ao et al. assembled a thermoelectric sensor by integrating the n-type Bi_2Te_3 flexible thin films with p-type Sb_2Te_3 counterparts and found that the thermal diffusion method is an effective way to fabricate high-performance, flexible Te-embedded Bi_2Te_3-based thin films [10]. With advancements in thermoelectric materials, thermoelectric power generation technology based on thermoelectric devices is expected to emerge as a new alternative energy technology.

There are several methods to improve the internal structure of a thermoelectric generator, such as optimizing the shape of the thermoelectric semiconductor, improving the thermal management scheme on the hot side of the TEM, and optimizing the TEG cooling system [11]. A pair of thermoelectric couples (TEC), the most fundamental constituent of a thermoelectric module, is often analyzed and optimized by researchers. The energy conversion efficiency and output performance of TEGs are affected by the length of the thermocouple legs, cross-sectional area, and spacing between the legs [12,13]. Chen et al. optimized the geometry of the TEC using a multi-objective genetic algorithm; the optimized output power and efficiency increased by about 51.9% and 5%, respectively [14]. Fan et al. investigated the effects of thermoelectric semiconductor leg cross-sectional area and length on power output, efficiency, and power density of the TEG under various thermal boundary conditions; they determined the optimal cross-sectional area ratio and length of thermocouple with the objective of maximizing peak output power [15]. To accommodate cylindrical heat sources, researchers proposed a new structure of annular thermoelectric couples (ATECs)—a number of ATECs were integrated and assembled into an annular thermoelectric generator (ATEG). The application of ATEG can effectively reduce the contact thermal resistance caused by a geometric mismatch between the cylindrical heat source and the flat-type TEG [16]. Zhu et al. studied the effect of ATEC geometric parameters on ATEG output power and efficiency in three application scenarios, determining the optimal shape factor under various boundary conditions [17]. Weng et al. investigated and improved variable angle ATEC geometry parameters, as well as designed a variable angle thermoelectric generator to increase output performance by 35% [18]. Furthermore, segmented ATEC was proposed [19–21], which effectively improved the thermoelectric performance of ATEG by taking the optimal operating temperature range of different thermoelectric materials into account.

Furthermore, improving the heat transfer performance between the hot fluid and hot end of TEG can significantly improve the overall performance of the thermoelectric system. Luo et al. designed a converging TEG with the hot side wall of the heat exchanger slanted inward, effectively increasing power output [22]. Li et al. placed foam metal with 20 pores per inch and a filling rate of 75% in a hot-side heat exchanger—the convective heat transfer coefficient of the channel was improved by a factor of four and the output power was doubled [23]. Yang et al. developed an ATEG based on a concentric annular heat exchanger for a cylindrical channel of an automotive exhaust pipe [24], which significantly improved the heat transfer coefficient and system net power.

The main cooling methods for thermoelectric devices in terms of waste heat dissipation at the cold end of TEG are heat sink heat dissipation [25,26], phase change material heat dissipation [27], air-cooling, and water-cooling [28–30]. Water-cooling and air-cooling methods, in particular, are widely used in TEG because of their superior cooling performance, as well as the advantages of a simple structure and broad applicability [31]. For example, He et al. proposed an optimized method to improve a flat plate TEG with ambient air-cooling [32]. Following that, He et al. studied the effect of different cooling methods on optimal TEG performance based on a common flat plate type TEG and found that the reverse flow of heat source and cooling fluid could achieve higher power output [33]. Luo et al. proposed a numerical model of an automotive TEG that used a flat-type tank as the cooling device and internally circulated water as the coolant to evaluate the TEG performance at different vehicle speeds [34].

The cooling methods of ATEG, however, have not been fully evaluated, although a series of investigations have been conducted on ATEG and its performance advantages over the flat type TEG demonstrated. Meanwhile, in practical applications, the exhaust temperature and exhaust mass flow rate vary with vehicle speed, resulting in changes in TEG temperature distribution and exhaust resistance. However, most studies concentrate solely on improving the performance of automotive TEG systems under constant operating conditions, ignoring the effects of vehicle speed and power loss due to exhaust resistance. Furthermore, the vast majority of the literature is based on conventional flat plate-type TEGs, and the applicability of these results and design guidelines to such ATEGs has not yet been verified, especially for the cooling method of ATEGs with a new concentric annular heat exchanger.

Thus, in this paper, a comprehensive numerical model of a thermo-fluid-electric multi-physics field for an ATEG with a concentric annular heat exchanger was developed. Temperature dependence of the physical properties of thermoelectric materials, heat transfer characteristics, effect of heat source parameters on exhaust resistance, and temperature gradient characteristics within the thermoelectric generator were considered. The effects of different heat source parameters on the heat transfer coefficient in the channel and the effects of different cooling methods on the optimal output power, net power, and energy conversion efficiency of this new TEG under different vehicle operating conditions were investigated. The new features of the ATEG thermal energy recovery system were also explored. The research findings may open up new avenues for the use of automotive exhaust heat recovery systems.

2. Mathematical Modeling of the CATEG

2.1. Three-dimensional Geometry of the CATEG

Figure 1a depicts a 3D schematic view of a water-cooling CATEG. The hot fluid inlet and outlet of the TEG are 45 mm in diameter, which matches the diameter of most automobile exhaust pipes. The ATEC are evenly distributed between the hot end heat exchanger and the radiator via an electrical series and thermal parallel connection. Figure 1b shows the axial profile of the CATEG. The concentric annular heat exchanger has a solid inner tube, and the automotive exhaust flows into the device at a temperature of T_{fin} and then flows through a narrow channel of the heat exchanger to heat the thermocouple, which then exits through the end of the heat exchanger. The proposed concentric annular heat exchanger compresses the fluid passages and improves heat transfer from the thermal fluid to the thermocouple, thereby improving the TEG output characteristics [24]. In the figure, L represents the length of TEG, while the inner and outer radiuses of the heat exchanger are represented by r_i and r_o, respectively, with r_i = 30 mm and r_o = 37 mm. Cooling water flows in at a temperature of T_{win} from inlets of the heat sink, effectively maintaining the temperature of the TEC cold end, and then flows out from the outlet at the other end. A portion of energy of the exhaust gas is transferred to the hot end as heat, which is directly converted to electrical energy by Seebeck effect of the thermoelectric elements, and the remainder is transferred to the cold end or to the outflow device.

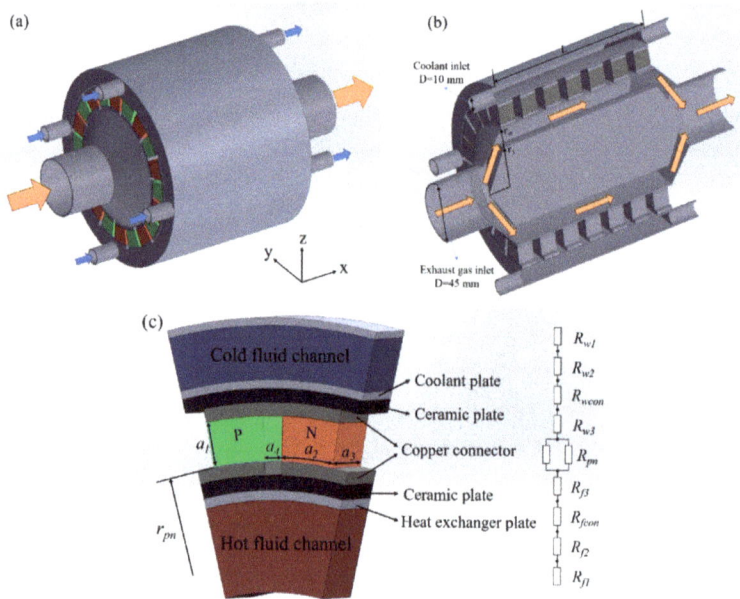

Figure 1. CATEG schematic view: (**a**) the whole frame, (**b**) profile of CATEG, (**c**) 3D view and equivalent thermal resistance of an ATEC.

Figure 1c shows the general structure of an annular thermoelectric couple branch and its equivalent thermal resistance. An annular thermoelectric couple is made up of an annular p-type thermoelectric semiconductor leg and an annular n-type thermoelectric semiconductor leg. Each PN leg is connected with a copper piece and the ATEC is insulated on the outside by two curved ceramic pieces (Al_2O_3). The thermocouple is made of commercially available Bi_2Te_3 material, which has variable resistivity, thermal conductivity, and Seebeck coefficient dependent on operating temperature. In the figure, a_1, a_2, and a_3 represent the height, inner arc length, and thickness of the ATEC, respectively, and a_4 represents the distance between the semiconductor legs. The inner radius of the PN leg is represented by r_{pn}; the thicknesses of ceramic sheets and copper sheets on both sides of the PN leg are denoted by δ_{cer} and δ_{cu}, respectively; and the wall thicknesses of the heat exchanger is denoted by δ_{plate}. Detailed parameters and material properties of the CATEG are shown in Table 1.

Table 1. Detailed parameters and material properties of the CATEG.

Parameters	Description	Value	Units
$a_1/a_2/a_3$	Height/thickness/inner arc length of the p(n)-type leg	5/5/5	mm
a_4	Distance between p-type leg and n-type leg	1	mm
δ_{cer}	Thickness of the ceramic sheet	0.05	mm
λ_{cer}	Thermal conductivity coefficient of ceramic	35	W m^{-1} K^{-1}
δ_{cu}	Thickness of the copper sheet	0.2	mm
λ_{cu}	Thermal conductivity coefficient of copper	398	W m^{-1} K^{-1}
δ_{plate}	Thickness of the exchanger plate	1	mm
λ_{plate}	Thermal conductivity coefficient of the exchanger plate	398	W m^{-1} K^{-1}
α_p	Seebeck coefficient of the p-type semiconductor	$\alpha_p(T) = 161 \times 10^{-4} - 1.818 \times 10^{-6}T + 1.11 \times 10^{-8}T^2 - 2.035 \times 10^{-11}T^3 + 1.134 \times 10^{-14}T^4$	V K^{-1}
α_n	Seebeck coefficient of the n-type semiconductor	$\alpha_n(T) = -4.428 \times 10^{-4} + 3.469 \times 10^{-6}T - 1.42 \times 10^{-8}T^2 + 2.325 \times 10^{-11}T^3 - 1.3 \times 10^{-14}T^4$	V K^{-1}
λ_p	Thermal conductivity of the p-type semiconductor	$\lambda_p(T) = -46.97 + 0.457T - 1.575 \times 10^{-3}T^2 + 2.331 \times 10^{-6}T^3 - 1.242 \times 10^{-9}T^4$	W m^{-1} K^{-1}
λ_n	Thermal conductivity of the n-type semiconductor	$\lambda_n(T) = 10.12 - 7.414 \times 10^{-2}T + 2.246 \times 10^{-4}T^2 - 3.019 \times 10^{-7}T^3 - 1.537 \times 10^{-10}T^4$	W m^{-1} K^{-1}
ρ_p	Electrical resistivity of the p-type semiconductor	$\rho_p(T) = -5.01 \times 10^{-5} + 3.519 \times 10^{-7}T - 7.74 \times 10^{-10}T^2 + 8.94 \times 10^{-13}T^3 - 4.32 \times 10^{-16}T^4$	Ω·m
ρ_n	Electrical resistivity of the n-type semiconductor	$\rho_n(T) = -8.072 \times 10^{-6} + 4.507 \times 10^{-8}T + 7.827 \times 10^{-11}T^2 - 2.305 \times 10^{-13}T^3 + 1.317 \times 10^{-16}T^4$	Ω·m

2.2. Main Equations of the Numerical Model

The ATEG non-isothermal finite element model is shown in Figure 2a. It can be divided into $n_x \times n_r$ pairs of thermocouples within the ATEG, with each pair of thermocouples acting as a computational unit. The i-th ring in the x-direction and the j-th thermocouple in the r-direction are chosen as an example, denoted as the (i, j)th computational unit, to illustrate the heat transfer process in this finite element model, where i ranges from 1 to n_x and j ranges from 1 to n_r. Following that, the CATEG modeling process was illustrated by using the example of cold and hot fluids flowing in the same direction, with the counterflow modeling process being similar to the cocurrent flow. Thermocouples installed in the same ring are connected in series, and ATECs in the same ring are assumed to have the same temperature distribution, thermodynamic properties, and power output; thus, the superscript i can be used to denote the inclusion of n_r pairs of ATECs. The numerical calculation is performed with each ring as a new calculation unit. The fluid temperature and ATEC surface temperature in the i-th ring are shown in Figure 2b. The automobile exhaust flows into the ATEG at temperature T_{fin} and the cold fluid flows into the device at temperature T_{win}. The hot fluid flows into the i-th ring at temperature T_f^i, and its heat is transferred to the ATEC hot end and ring $i + 1$, respectively, to raise the hot end temperature of the thermocouple to T_h^i. The hot fluid flows out of ring i at temperature T_f^{i+1}. Similarly, the cold fluid flows into the i-th ring at temperature T_w^i and its heat is transferred to the ATEC cold end and the $i + 1$th ring, respectively, cooling the ATEC cold end to T_c^i and then flowing out of the i-th ring at temperature T_w^{i+1}. The precondition in this model is that there should be no air in the thermoelectric semiconductors, and the Thomson effect and thermal radiation could be ignored.

Three sets of heat transmission equations can be used to describe the heat transfer rate at the hot side of the ATEG, Q_h, and the heat transfer rate at the cold side, Q_c [35]. The first group is represented by the components of the Peltier effect, conduction heat, and Joule heat transfer to both ends of the TEC, respectively; the second group is constructed by considering the rate of heat transfer to the fluid; and the third group is the heat transferred by convection to the fluid at the hot and cold sides of the solid phase given by the cooling Newton law, respectively.

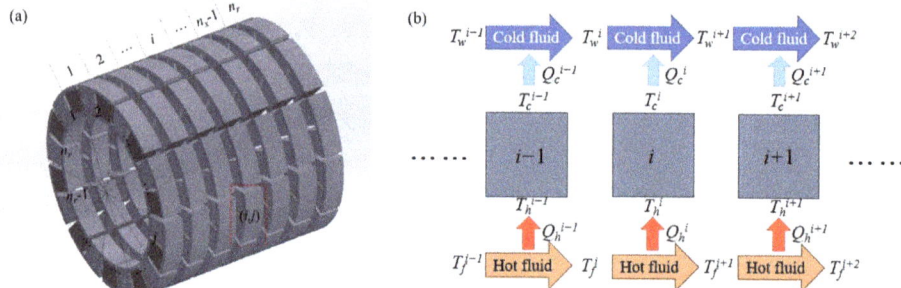

Figure 2. (a) Finite element analysis of the annular thermoelectric generator, (b) schematic of calculation unit with temperature definitions (cocurrent flow).

Based on the non-isothermal finite element model and steady-state heat transfer process in a single ring, the two heat transmission components $Q_h{}^i$ and $Q_c{}^i$ are described as:

$$\begin{cases} Q_h^i = n_r[\alpha_{pn}{}^i I T_h^i + K_{pn}{}^i(T_h^i - T_c^i) - 0.5I^2 R_{pn}{}^i] \\ Q_c^i = n_r[\alpha_{pn}{}^i I T_c^i + K_{pn}{}^i(T_h^i - T_c^i) + 0.5I^2 R_{pn}{}^i] \end{cases} \quad (1)$$

The second group, $Q_{transff}$ and $Q_{transfw}$, are given by:

$$\begin{cases} Q_{transff}^i = c_f m_f (T_f^i - T_f^{i+1}) \\ Q_{transfw}^i = c_w m_w (T_w^{i+1} - T_w^i) \end{cases} \quad (2)$$

The third group, Q_{convf} and Q_{convw}, are given by:

$$\begin{cases} Q_{convf}^i = n_r A_h k_f [0.5(T_f^i + T_f^{i+1}) - T_h^i] \\ Q_{convw}^i = n_r A_c k_w [T_c^i - 0.5(T_w^{i+1} + T_w^i)] \end{cases} \quad (3)$$

where A denotes the heat transmission area of the thermocouple, the subscripts "h" and "c" denote the hot end and cool end, respectively; c and m denote the specific heat capacity and mass flow rate, respectively; k denotes the total heat transfer coefficient of the fluid; and the subscripts "f" and "w" denote the hot and cold fluids, respectively.

According to the continuity condition at the junctions, we have $Q_h = Q_{convf}$, $Q_c = Q_{convw}$. The heat absorbed by the ATEG hot-side junction is equal to the heat released by the hot fluid, and the heat released by the TEG cold-side junction is equal to the heat absorbed by the cold fluid. Therefore, we have $Q_h = Q_{transff}$, $Q_c = Q_{transfw}$.

The Seebeck coefficient α_{pn}, thermal conductance K_{pn}, and resistance R_{pn} of an ATEC can be calculated using the following equations, respectively:

$$\alpha_{pn}{}^i = \overline{\alpha_p{}^i} - \overline{\alpha_n{}^i} \quad (4)$$

$$K_{pn}{}^i = a_2 a_3 (\overline{\lambda_p{}^i} + \overline{\lambda_n{}^i}) / \{r_{pn} \ln[(r_{pn} + a_1)/r_{pn}]\} \quad (5)$$

$$R_{pn}{}^i = r_{pn} \ln[(r_{pn} + a_1)/r_{pn}](\overline{\rho_p{}^i} + \overline{\rho_n{}^i})/a_2 a_3 \quad (6)$$

The temperature-dependent equations for α, ρ, and λ of thermoelectric material are determined by:

$$\begin{aligned} \overline{\alpha_p{}^i} &= [\int_{T_c^i}^{T_h^i} \alpha_p(T) dT]/[T_h{}^i - T_c{}^i] \\ \overline{\alpha_n{}^i} &= [\int_{T_c^i}^{T_h^i} \alpha_n(T) dT]/[T_h{}^i - T_c{}^i] \end{aligned} \quad (7)$$

$$\overline{\lambda_p^i} = [\int_{T_c^i}^{T_h^i} \lambda_p(T)dT]/[T_h^i - T_c^i]$$
$$\overline{\lambda_n^i} = [\int_{T_c^i}^{T_h^i} \lambda_n(T)dT]/[T_h^i - T_c^i] \quad (8)$$

$$\overline{\rho_p^i} = [\int_{T_c^i}^{T_h^i} \rho_p(T)dT]/[T_h^i - T_c^i]$$
$$\overline{\rho_n^i} = [\int_{T_c^i}^{T_h^i} \rho_n(T)dT]/[T_h^i - T_c^i] \quad (9)$$

As shown in Figure 1c, the thermal resistance in the process of fluid heat transfer primarily consists of the convective thermal resistance R_{f1} of the thermal fluid, the thermal conductivity R_{f2} through the heat exchanger, the contact thermal resistance R_{fcon} between the heat exchanger and the ceramic sheet, and the conduction thermal resistance R_{f3} through the ceramic piece and the copper connector; similarly, the cold end thermal resistance is R_{w1}, R_{w2}, R_{wcon}, and R_{w3}. Therefore, the total heat transfer coefficient k_f is determined by:

$$\begin{aligned} k_f &= 1/(R_1 + R_2 + R_{con} + R_3) \\ &= 1/(1/h + \delta_{plate}/\lambda_{plate} + R_{fcon} + \delta_{cu}/\lambda_{cu} + \delta_{cer}/\lambda_{cer}) \end{aligned} \quad (10)$$

where h denotes the convective heat transfer coefficient, and is given by

$$h = Nu\lambda_f/D \quad (11)$$

where D denotes the hydraulic diameter of the hot fluid channel. In (11), the Nusselt number Nu of hot fluid is determined by Gnielinski-related estimation [22]:

$$Nu = \begin{aligned} &0.0214(Re^{0.8} - 100)Pr^{0.4}[1+ \\ &(D_h/L)^{2/3}](T_{fav}/T_{wav})^{0.45}, 2300 \le Re \le 10^6 \end{aligned} \quad (12)$$

where T_{fav} and T_{wav} denote the average temperature at which the hot and cold fluids flow through the TEG; Pr and Re represent the Prandtl number and Reynolds number of the hot fluid, respectively; Re can be calculated as follows:

$$Re = \gamma_f D v_f / \mu_f \quad (13)$$

where γ, μ, and v are the density, dynamic viscosity, and velocity, respectively.

The output performances of the CATEG, i.e., total current, power output, and efficiency equations are as follows:

$$I = \sum_{i=1}^{n_x} \alpha_{pn}(T_h^i - T_c^i)/(R_L + R_{pn}(i)n_x n_r) \quad (14)$$

$$P_{teg} = \sum_{i=1}^{n_x} (Q_h^i - Q_c^i) \quad (15)$$

$$\eta = P_{teg}/\sum_{i=1}^{n_x} Q_h^i \quad (16)$$

Although the heat transfer is improved in a CATEG, the reduction in hydraulic diameter results in a larger pressure drop. As a result, an evaluation of the CATEG net power is required. The pressure drop Δp of the channel is expressed as:

$$\Delta p = 4F(L/D)(v_f^2 \rho_f/2) \quad (17)$$

where Darcy resistance coefficient F is defined as follows [36]:

$$\begin{cases} F = 0.0791/Re^{0.25}, 2000 < Re \leq 59.7/(2H_r/D)^{8/7} \\ 0.5/\sqrt{F} = -1.8\lg\left\{6.8/Re + (H_r/3.7D)^{1.11}\right\}, \\ 59.7/(2H_r/D)^{8/7} < Re \leq 665 - 765\lg(2H_r/D)/(2H_r/D) \\ F = 0.25/\{2\lg[3.7D_h/(2H_r/D)]\}^2, \\ Re > 665 - 765\lg(2H_r/D)/(2H_r/D) \end{cases} \quad (18)$$

where H_r = 0.005 mm denotes the surface finish quality of the concentric annular heat exchanger.

The power dissipation P_b caused by the exhaust back pressure is calculated by combining Equations (13), (17) and (18):

$$P_b = \Delta p(m_f/\rho_f) \quad (19)$$

The net power P_{net} of the CATEG system is calculated as:

$$P_{net} = P_{teg} - P_b \quad (20)$$

2.3. Solution Method

To solve the temperature distribution and heat distribution of ATEG using the numerical model developed in Section 2.2, the following equations were constructed from Equations (1)–(3):

$$\begin{bmatrix} 1/c_f m_f & 0 & 0 & 0 & 1 & 0 \\ 1/n_r & 0 & -(\alpha_{pn}{}^i I + K_{pn}{}^i) & K_{pn}{}^i & 0 & 0 \\ 1/n_r A_h k_f & 0 & 1 & 0 & -0.5 & 0 \\ 0 & 1/c_w m_w & 0 & 0 & 0 & -1 \\ 0 & 1/n_r & -K_{pn}{}^i & K_{pn}{}^i - \alpha_{pn}{}^i I & 0 & 0 \\ 0 & 1/n_r A_c k_w & 0 & -1 & 0 & 0.5 \end{bmatrix} \begin{bmatrix} Q_h^i \\ Q_c^i \\ T_h^i \\ T_c^i \\ T_f^{i+1} \\ T_w^{i+1} \end{bmatrix} = \begin{bmatrix} T_f^i \\ -0.5I^2 R_{pn}{}^i \\ 0.5T_f^i \\ -T_w^i \\ 0.5I^2 R_{pn}{}^i \\ -0.5T_w^i \end{bmatrix} \quad (21)$$

The total series current I of the CATEG is highly algebraically coupled to the nonlinear system of Equation (21), i.e., the temperature distribution of CATEG affects the series current I, and I affects the temperature distribution inside the thermoelectric generator at the same time; no analytical solution was obtained. Similar coupling relations exist between the TEG temperature field and the total heat transfer coefficient. A double iterative circular approximation approach was used to effectively solve such a multiple coupling numerical problem and the entire calculation process was solved using the Matlab program. The flowchart of the CATEG model solution procedure is shown in Figure 3.

First, the model parameters were provided and the initial guess I_0 and k_{f0} initialization procedures were set. The initial temperature distribution within the CATEG was obtained by solving Equation (21). Based on this temperature distribution, Equation (10) was used to update k_f, which was then used as the new k_{f0} for the next inner loop iteration. The process was repeated until the k_f equaled the new initial k_{f0} of this iteration. Next, the new series current I was calculated using Equation (14) and used as the initial current I_0 in the next iteration. This process was repeated until the new series current I equaled the initial current I_0 of this iteration. Once the current, total heat transfer coefficient, and temperature distribution were determined, the power output, efficiency, and net power of CATEG was calculated using (15), (16), and (20).

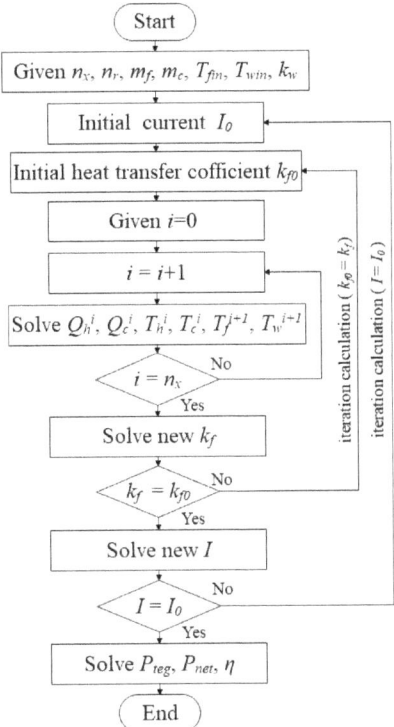

Figure 3. Flowchart of the CATEG model solution procedure.

3. Model Validation

In this section, the numerical results of the developed thermo-fluid-electric multi-physics field CATEG system model are validated with other numerical models and experimental data.

The numerical results of this model are compared with those of Yang et al. [37], who developed a numerical model of an ATEG applicable to a cylindrical heat source to comprehensively evaluate the optimal thermoelectric material structure size and maximum net power. To reproduce their data, we used the same ATEG parameter settings as in the literature in the model validation and set the hot end heat exchanger to a common cylindrical channel. Figure 4a depicts the variation in the ATEG maximum net power with hot fluid mass flow rate. The numerical simulation results in this study have a maximum error of 4.3% with Ref. [37]. This error is due to the fact that our model considers heat conduction through the ceramic and copper sheets, as well as the contact thermal resistance between the thermoelectric semiconductor and the heat exchanger, which were not considered in Ref. [37]. Furthermore, we improved the calculation method of the physical properties of the thermocouple, i.e., the Seebeck coefficient, thermal conductivity, and resistivity of the thermoelectric semiconductor are all highly dependent on temperature.

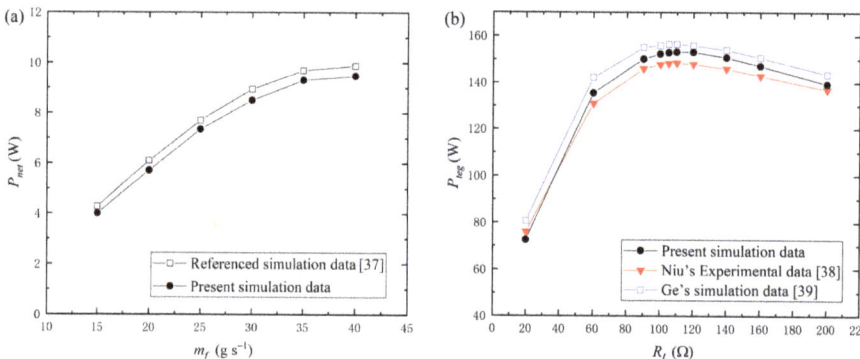

Figure 4. Validation of the proposed ATEG numerical model. (**a**) Comparison with simulation data in [37] (Adapted with permission from Ref. [37]. 2019 Elsevier). (**b**) Comparison with experimental data in [38] (Adapted with permission from Ref. [38]. 2008 Elsevier) and simulation data in [39] (Adapted with permission from Ref. [39]. 2019 Elsevier).

The numerical simulation results of this paper were compared with the experimental results as well as the numerical results of Ge et al. to further validate the developed CATEG numerical model. Niu et al. built an experimental thermoelectric generator set out of 56 commercially available Bi_2Te_3 thermoelectric modules combined with a flat plate heat exchanger and tested the power generation performance at various temperature differences [38]. Ge et al. proposed a new annular thermoelectric vaporizer that combined an air-heated vaporizer and thermoelectric power generation technology [39], and compared the simulation results with Niu's experimental data. This is because, when the radius of the annular thermocouple is large enough, the inner arc of the thermoelectric leg corresponds to a small curvature and can be considered as a flat plate type thermocouple. The parameters from Ref. [38] were used for model verification. Figure 4b depicts the output performance of the thermoelectric generator under different loads. The error between the simulation data of Ge and the experimental results is less than 10%, whereas the error between the numerical simulation results in this study and the experimental data is around 7%. Although the numerical model established in this paper is a general thermoelectric generator mechanism model, there are still some discrepancies between simulation results and experimental data, which are primarily due to the model parameter settings limitations. First, the contact thermal resistance is set to 0.0008 $m^2 \cdot K\ W^{-1}$ in the simulation; however, this is not provided in the experiment. Second, the density, dynamic viscosity, and Prandtl number of the heat source are highly dependent on temperature, which is not measured in the experiment but is set to a constant value in this model.

4. Results and Discussion

4.1. Effect of Different Cooling Methods on CATEG Thermoelectric Conversion Performance

The type of cooling fluid used in CATEGs for an automobile waste heat recovery system can be divided into air-cooling (ambient air used as cooling fluid during automobile operation) and water-cooling (coolant of the automobile engine cooling system). In regards to fluid flow direction, it can be divided into two ways: hot and cold fluid flowing in the same and reverse directions [33]. As a result, when the exhaust direction is fixed, the four most common cooling methods for an automobile thermoelectric generator system are cocurrent water-cooling (COW), cocurrent air-cooling (COA), countercurrent water-cooling (COUW), and countercurrent air-cooling (COUA). Since the convective heat transmission thermal resistance between the thermocouple and the ambient air (or water) is significantly lower than that between the thermocouple and the hot fluid [40], the heat transfer coefficient at the cold end is assumed to be constant. The cold side heat transfer coefficient k_w is

100 W m^{-2} K^{-1} for air-cooling and 1000 W m^{-2} K^{-1} for water-cooling. Table 2 shows the specific parameters of the hot and cold fluids.

Table 2. Detailed parameters and properties of the hot and cold fluids.

Name	Description	Parameter	Value	Unit
Exhaust gas	Heat transfer coefficient	k_f	Equation (10)	W m^{-2} K^{-1}
	Inlet temperature	T_{fin}	400	°C
	Mass flow rate	m_f	20	g s^{-1}
	Specific heat capacity	c_f	1.12	J g^{-1} K^{-1}
Ambient air	Heat transfer coefficient	k_w	100	W m^{-2} K^{-1}
	Inlet temperature	T_{win}	30	°C
	Mass flow rate	m_w	20	g s^{-1}
	Specific heat capacity	c_w	1.0	J g^{-1} K^{-1}
Cooling water	Heat transfer coefficient	k_w	1000	W m^{-2} K^{-1}
	Inlet temperature	T_{win}	70	°C
	Mass flow rate	m_w	200	g s^{-1}
	Specific heat capacity	c_w	4.177	J g^{-1} K^{-1}

The variation in the power output and energy conversion efficiency of CATEG with the volume of thermoelectric semiconductor for different cooling methods is shown in Figures 5 and 6, respectively. When compared to the cocurrent cooling method, the countercurrent cooling method increases the output power and efficiency of the thermoelectric generator; the percentage increase is also shown in the figures. This shows that the improved output power and efficiency by countercurrent water-cooling are lower than those improved by countercurrent air-cooling; when the total thermocouple volume is less than 2.66×10^{-4} m^3, the increased percentage of output power and efficiency is lower and nearly the same for both cocurrent and countercurrent.

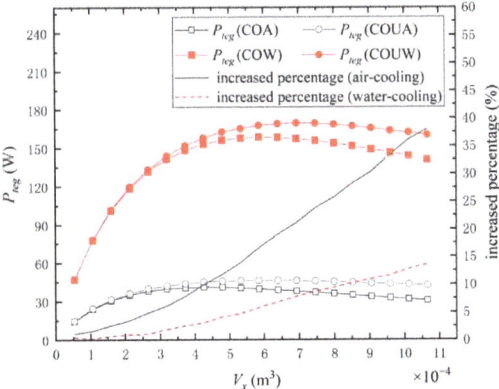

Figure 5. Variations in CATEG output power with the total thermoelectric semiconductor volume under different cooling methods and the increased percentage obtained by the countercurrent flow method over that obtained by the cocurrent flow method.

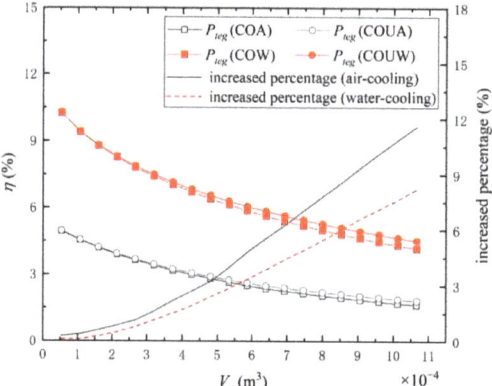

Figure 6. Variations in CATEG energy conversion efficiency with the total thermoelectric semiconductor volume under different cooling methods and the increased percentage obtained by the countercurrent flow method over that obtained by the cocurrent flow method.

This is due to the fact that when the total thermocouple volume is small, the length of the CATEG is short, and cooling water flow direction has little influence on the temperature change at the TEG cold end; when the volume of the thermocouple increases, the use of countercurrent water-cooling can optimize the temperature distribution of the thermoelectric generator, i.e., obtain a more stable temperature difference to improve the output performance of the CATEG. Simultaneously, when using countercurrent air-cooling, the power output and energy conversion efficiency of the TEG are significantly improved as the thermocouple volume gradually increases, compared to cocurrent air-cooling. This is because the flow direction of air has a large impact on the cooling performance of the TEG: the mass flow rate of air is small compared to the cooling water, and the working temperature difference of the thermocouple decreases continuously when the exhaust gas and air flow in the same direction, whereas the countercurrent flow method can effectively maintain the stability of the working temperature difference of the ATECs and improve the thermoelectric conversion performance of the TEG. The greater the total volume of the thermocouple, the greater the output power and efficiency boosted by the counterflow of cold and hot fluids, and the percentage boosted by air-cooling is always greater than the percentage boosted by water-cooling. Furthermore, for each of the four cooling methods, there is an optimal thermocouple volume that produces the highest output power of the thermoelectric generator, and in general, the optimal volume is larger for water-cooling than air-cooling, and larger for countercurrent flow than cocurrent flow.

The variation in the working temperature difference ΔT of the thermocouple for the four different cooling methods are shown in Figure 7. The effect of the cocurrent flow and countercurrent flow methods on ΔT is small for water-cooling, but significant for air cooling. When using cocurrent air-cooling, ΔT decreases rapidly along the direction of fluid flow, whereas when using countercurrent air-cooling, the thermal energy transferred from the hot fluid to the cold side of the thermocouple gradually heats the air, resulting in a decreasing ΔT along the direction of air inflow; however, in general, the countercurrent flow method improves the operating temperature difference of the thermocouples and keeps the TEG at a higher output performance. When using cocurrent water-cooling, ΔT decreases rapidly along the direction of fluid flow, similar to the case of ACO; when using countercurrent water-cooling, ΔT also decreases continuously along the direction of exhaust gas flow, but at a faster rate than cocurrent flow. This is actually determined by the physical properties of hot and cold fluids, because the mass flow rate as well as the specific heat capacity of water are much greater than that of exhaust gas, regardless of whether the water is cocurrent or countercurrent flowing through the TEG, the temperature change is not too large, and thus has similar thermoelectric performance. Although the maximum temperature difference of COW is greater than that of COUW, the average working temperature difference of thermocouples in COUW is slightly greater than that of COW, and this difference is not obvious in practical applications.

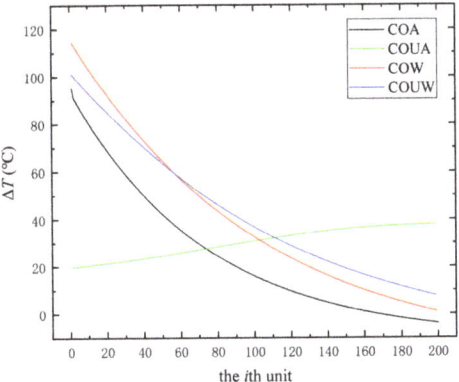

Figure 7. Variations in the working temperature difference ΔT of the thermocouple for the four different cooling methods.

As a result, when using air-cooling in the CATEG system, a distinction must be made between using the cocurrent flow or countercurrent flow method depending on the actual situation, whereas water-cooling does not require a distinction because the effect on thermoelectric properties is minor. On the other hand, the effect of different cooling methods on the hot side heat transfer coefficient within the CATEG can demonstrate this result. Figure 8 depicts the variations in total heat transfer coefficient at the hot end of the CATEG for different cooling methods, as well as the increased percentage obtained by the countercurrent flow method over that obtained by the cocurrent flow method. As shown, the hot side heat transfer coefficient decreases continuously as the thermoelectric semiconductor volume increases under the four cooling methods. The change in cooling fluid flow direction has very little effect on the heat transfer coefficient; the real reason for improved thermoelectric generator performance is that the countercurrent flow method increases the working temperature difference of the thermocouples, which increases efficiency and output power.

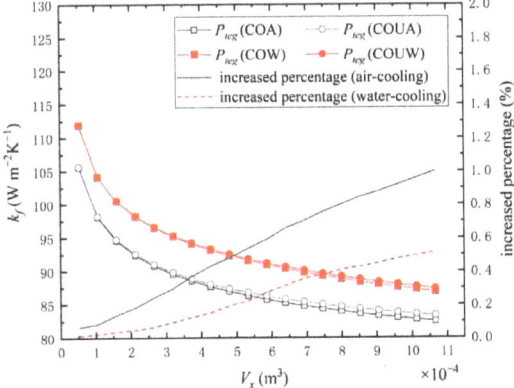

Figure 8. Variations in the total heat transfer coefficient with the total volume of the thermoelectric semiconductor for different cooling methods.

4.2. Effect of Various Exhaust Parameters on CATEG Performance under Four Different Cooling Methods

Section 4.1 discussed and analyzed the variation in CATEG output performance with total volume of thermoelectric semiconductors for different cooling methods; however, changes in vehicle operating conditions cause changes in exhaust parameters, which also have a significant impact on the thermoelectric performance of CATEG, and changes in engine exhaust back pressure caused by exhaust parameters' fluctuation in a CATEG can also result in additional engine power loss. Therefore, it is critical to conduct an analysis of the effects of various automobile operating conditions on CATEG performance. According to Ref. [35], during vehicle operation, the exhaust gas temperature and mass flow rate vary with vehicle operating conditions within 200–600 °C and 10–50 g s^{-1}, respectively, and this section evaluates the output performance of CATEG for various operating conditions and cooling methods.

The variation in CATEG power output for various exhaust gas temperatures under different cooling methods are shown in Figure 9. From Figure 9a,b, for the air-cooling method, the power output of the countercurrent method increases with the increase in thermoelectric semiconductor volume and then slowly decreases, and its maximum power output is significantly higher than that of the cocurrent flow method; the advantage of the countercurrent method becomes more apparent as the inlet temperature rises. Meanwhile, the optimal thermoelectric semiconductor volume corresponding to the maximum power output point is greater than that of the cocurrent flow method. The power output increases and then decreases as the thermoelectric semiconductor volume increases under the countercurrent flow method, and CATEG maintains a high output power for only a limited range of thermoelectric semiconductor volume. The optimal total thermoelectric volume is defined as the volume corresponding to the maximum output power point. The optimal total thermoelectric semiconductor volume for the countercurrent air-cooling (9.06×10^{-4} m^3 for P_{net} = 57.94 W) is greater than that of the cocurrent air-cooling (4.26×10^{-4} m^3 for P_{net} = 45.7 W). When using cocurrent air-cooling, the power output increases and then decreases as the heat source temperature rises, in contrast to the other three methods, which increase with increasing heat source temperature. Because of the presence of a large temperature gradient, the Seebeck coefficient and resistance of the PN couples along the direction of fluid flow continue to decrease during TEG operation, while heat transfer through the thermocouple increases. This temperature gradient characteristic becomes more apparent when the exhaust inlet temperature is higher. Although increasing the inlet temperature raises the working temperature difference of the thermocouple, which increases the output power, this power in the cocurrent air-cooling mode is insufficient to compensate for the power consumed by the thermoelectric semiconductors, resulting in a reduced output power. Figure 9c,d show that when water-cooling is used, the effect of the cocurrent and countercurrent methods on TEG thermoelectric performance is small; and the effect of exhaust inlet temperature on thermoelectric performance is similar under both cooling methods. The countercurrent flow method has a 5.5% higher maximum output power than the cocurrent flow method, which is consistent with the discussion in Section 4.1.

Figure 10 shows the variation in CATEG power output for various exhaust mass flow rates under different cooling methods. As shown in Figure 10a,b, the CATEG power output increases and then decreases with the rise in thermoelectric semiconductor volume for different exhaust gas mass flow rates under the air-cooling method; when the volume of the thermoelectric semiconductor remains constant, the output power increases quickly at first with increasing mass flow rate and then gradually decreases. When m_f = 40 g s^{-1}, the maximum power output of these two cooling methods can be obtained separately, but the maximum power point in the countercurrent method corresponds to a thermoelectric semiconductor volume of $V_x = 5.86 \times 10^{-4}$ m^3, which is larger than that in the cocurrent flow method $V_x = 4.26 \times 10^{-4}$ m^3, and the maximum output power of the countercurrent method is 13.2% higher than that of the cocurrent method.

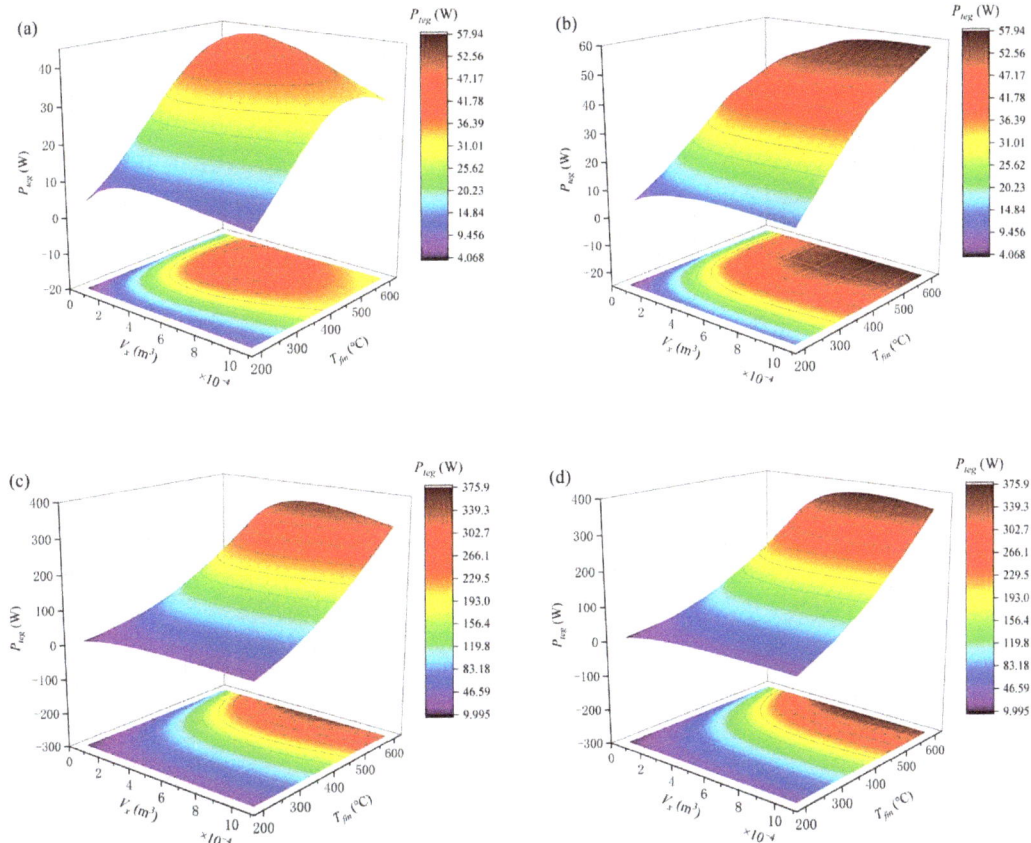

Figure 9. Variations in CATEG power output for various exhaust gas temperatures under different cooling methods when m_f = 20 g s^{-1}: (**a**) cocurrent air-cooling, (**b**) countercurrent air-cooling, (**c**) cocurrent water-cooling, (**d**) countercurrent water-cooling.

According to Figure 10c,d, when using water-cooling, the power output increases first with increase in the volume of the thermoelectric semiconductor, and once the volume reaches a certain value, effective power boost cannot be obtained by further adding the thermoelectric element. The output power variation in the COW and COUW methods is nearly identical, but the maximum output power of the countercurrent flow is significantly greater than that of the cocurrent flow, especially at high m_f (more than 20 g s^{-1}). The mass flow rate of the exhaust gas has a significant influence on the electricity output of a water-cooling CATEG. When the volume of the thermoelectric semiconductor remains constant, the output power increases almost linearly as the mass flow rate increases. The optimal total thermoelectric semiconductor volume for the countercurrent water-cooling (1.06 × 10^{-3} m^3 for P_{net} = 515.9 W) is similar to that of the cocurrent water-cooling (9.59 × 10^{-4} m^3 for P_{net} = 466.08 W), and the maximum output power of the countercurrent method is 10.6% greater than that of the cocurrent method. The output power of the water-cooling is much higher compared to the air-cooling, regardless of fluid flow direction, especially at large mass flow rates.

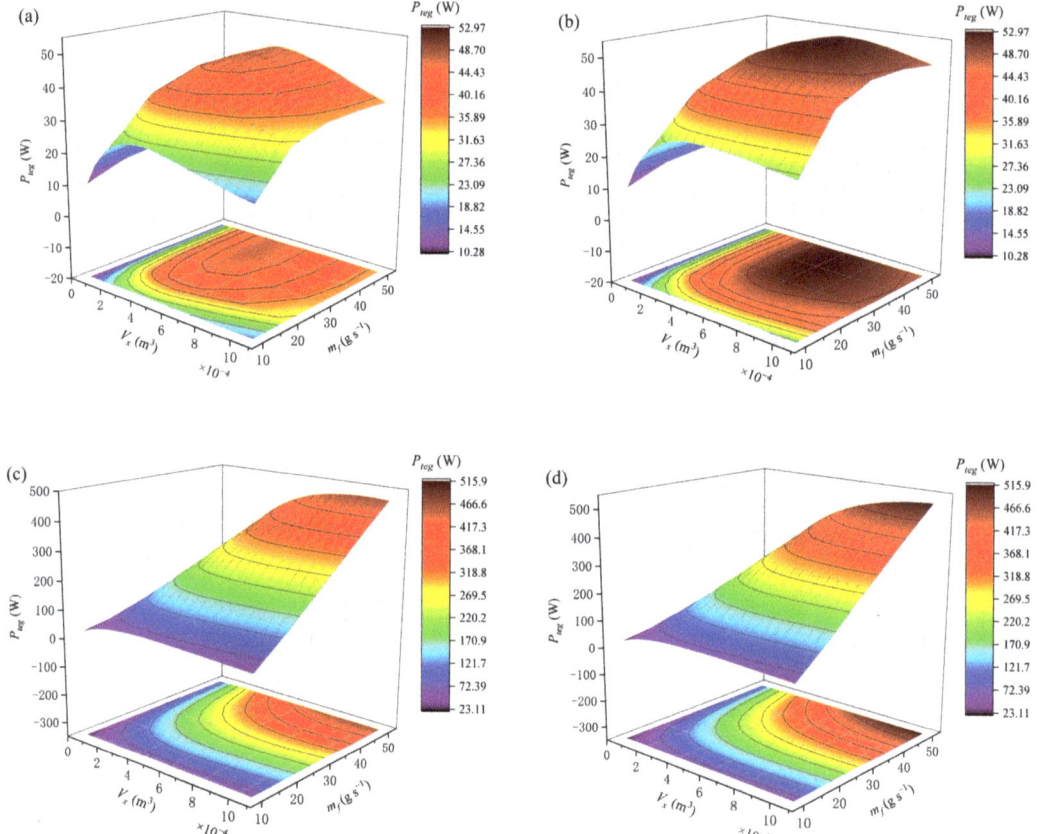

Figure 10. Variations in CATEG power output for various exhaust gas mass flow rates under different cooling methods when T_{fin} = 400 °C: (**a**) cocurrent air-cooling, (**b**) countercurrent air-cooling, (**c**) cocurrent water-cooling, (**d**) countercurrent water-cooling.

4.3. Effect of Exhaust Mass Flow Rates on CATEG Net Power under Four Different Cooling Methods

Although the output power of the countercurrent flow cooling method is higher than that of the cocurrent flow method, as previously discussed, a larger thermoelectric semiconductor volume is required to achieve optimal thermoelectric performance, implying a longer TEG length and a larger device volume, which will lead to an increase in the exhaust back pressure of the car engine, resulting in a decrease in net system power. Therefore, the effect of exhaust mass flow rate on net power of the CATEG system under different cooling methods will be investigated in this section.

Figure 11 shows the variation in CATEG net power for various exhaust gas mass flow rates under different cooling methods. The net power of the air-cooling CATEG increases and then decreases as the thermoelectric semiconductor volume increases, as shown in Figure 11a,b. The output power of the TEG at high mass flow rates is insufficient to compensate for the power loss to the system caused by high exhaust back pressure, resulting in a negative net power. The exhaust mass flow rate has a significant effect on the net power of CATEG, when the volume of the thermoelectric semiconductor remains constant; the net power increases with increasing mass flow rate and then rapidly decreases. When m_f =20 g s^{-1}, the maximum net power of these two cooling methods can be obtained separately; the maximum net power point in COUA method corresponds to a thermoelectric semiconductor volume of $V_x = 5.33 \times 10^{-4}$ m^3, while in COA, its $V_x = 4.26 \times 10^{-4}$ m^3; the maximum net power of the countercurrent method is 11.2% greater than that of the cocurrent method.

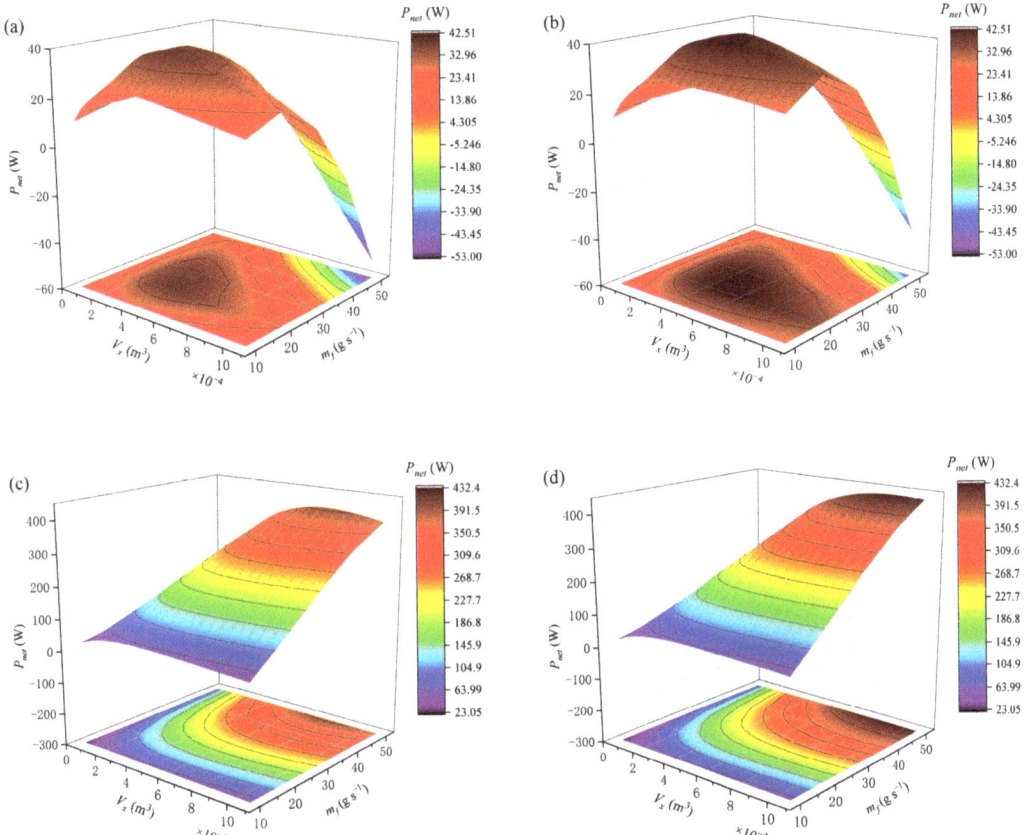

Figure 11. Variations in CATEG net power for various exhaust gas mass flow rates under different cooling methods when $T_{fin} = 400$ °C: (**a**) cocurrent air-cooling, (**b**) countercurrent air-cooling, (**c**) cocurrent water-cooling, (**d**) countercurrent water-cooling.

From Figure 11c,d, it can be seen that countercurrent flow or cocurrent flow has little effect on the net power of water-cooling in comparison to air-cooling. The optimal total thermoelectric semiconductor volume for the countercurrent water-cooling (9.06×10^{-4} m^3 for $P_{net} = 432.42$ W) is greater than that of the cocurrent water-cooling (7.46×10^{-4} m^3 for $P_{net} = 396.87$ W). Although the maximum net power is increased by 8.9%, the optimal thermoelectric semiconductor volume is increased by 21.4%; therefore, the economics of CATEG is slightly reduced. When the goal is to maximize CATEG net power, the determined optimal thermoelectric semiconductor volume is very different from that determined in Section 4.2; therefore, when designing and improving automobile exhaust TEG and choosing cooling methods, the impact of power loss caused by the engine exhaust back pressure on the TEG system should be considered.

5. Conclusions

In this study, a comprehensive variable physical property numerical model of CATEG using the finite element method was developed. The established theoretical model was validated using experimental data and other numerical simulation results. The maximum output power, net power, and optimal total thermoelectric semiconductor volume were calculated and compared for four different cooling methods with different exhaust parameters. The primary findings of the study are summarized below:

(1) When compared to the cocurrent cooling method, the countercurrent cooling method can effectively improve the working temperature difference of the thermocouples, especially when using air-cooling, thereby increasing the output power; however, it requires more thermoelectric semiconductor volume to achieve maximum output power;

(2) It is not preferable to use the cocurrent air-cooling method for the heat source with high temperature. As the temperature of the heat source rises, the output power of TEG increases first, then gradually declines after reaching a peak. For COUA, COW, and COUW, the output power rises almost linearly as the temperature of the heat source increases;

(3) The exhaust mass flow rate has a significant influence on CATEG net power. The maximum net power P_{net} = 432.42 W can be obtained using countercurrent water-cooling, corresponding to an optimal thermoelectric semiconductor volume of 9.06×10^{-4} m^3. Compared to COW, the maximum net power increased by 8.9%, but the optimal thermoelectric semiconductor volume increased by 21.4%.

Author Contributions: Data curation, writing—original draft preparation, W.Y.; formal analysis, methodology, W.Z.; conceptualization, funding acquisition, writing—review and editing, Y.Y. and C.X.; validation, L.H.; visualization, Y.S. All authors have read and agreed to the published version of the manuscript.

Funding: This research was funded by the National Natural Science Foundation of China (51977164).

Institutional Review Board Statement: Not applicable.

Informed Consent Statement: Not applicable.

Data Availability Statement: All data, models, or code that support the findings of this study are available from the corresponding author upon reasonable request. The data are not publicly available as the data also forms part of an ongoing study.

Conflicts of Interest: The authors declare no conflict of interest.

Nomenclature

a_1, a_2, a_3	height, inner arc length, and thickness of the thermoelectric leg, mm
a_4	gap between p- and n-type semiconductors, mm
C	specific heat capacity, J·g^{-1}·K^{-1}
D	diameter, mm
D_h	hydraulic diameter, m
F	Darcy resistance coefficient
H_r	surface finish quality, m
H	convective heat transfer coefficient, W·m^{-2}·K^{-1}
I	current, A
K	total heat transfer coefficient, W·m^{-2}·K^{-1}
K	thermal conductance, W·K^{-1}
L	length of the heat exchanger, m
m	mass flow rate, g·s^{-1}
n_r	total thermocouple number in a single-ring
n_x	total thermocouple number in a line
Nu	Nusselt number
p	pressure, Pa
P	power, W
Pr	Prandtl number

Q	quantity of heat, W
r	radius, m
R	resistance, Ω
Re	Reynolds number
T	temperature, °C

Greek symbols

α	the Seebeck coefficient, $V \cdot K^{-1}$
γ	density, $kg \cdot m^{-3}$
δ	thickness, mm
Δ	difference
η	efficiency, %
λ	thermal conductivity, $W \cdot m^{-1} \cdot K^{-1}$
μ	dynamic viscosity, $Pa \cdot s$
ρ	resistivity, $\Omega \cdot m$

Subscript

b	consumed pump value
c	cold side of the thermoelectric generator
cer	ceramic
con	connector
cu	copper
f	hot fluid
fav	average value of hot fluid
h	hot side of the thermoelectric generator
i	inner ring of the hot end heat exchanger
L	external load
n	n-type thermoelectric semiconductor
net	net value
o	outer ring of heat exchanger
p	p-type thermoelectric semiconductor
$plate$	heat exchanger plate
pn	thermocouple
teg	TEG system value
w	cold fluid
wav	average value of cold fluid

Abbreviations

ATEC	annular thermoelectric couple
ATEG	annular thermoelectric generator
CATEG	concentric annular thermoelectric generator
TEG	thermoelectric generator

References

1. Li, X.; Xie, C.; Quan, S.; Huang, L.; Fang, W. Energy management strategy of thermoelectric generation for localized air conditioners in commercial vehicles based on 48 V electrical system. *Appl. Energy* **2018**, *231*, 887–900. [CrossRef]
2. Zhang, Z.; Zhang, Y.; Sui, X.; Li, W.; Xu, D. Performance of Thermoelectric Power-Generation System for Sufficient Recovery and Reuse of Heat Accumulated at Cold Side of TEG with Water-Cooling Energy Exchange Circuit. *Energies* **2020**, *13*, 5542. [CrossRef]
3. Musharavati, F.; Khanmohammadi, S.; Nondy, J.; Gogoi, T.K. Proposal of a new low-temperature thermodynamic cycle: 3E analysis and optimization of a solar pond integrated with fuel cell and thermoelectric generator. *J. Clean. Prod.* **2022**, *331*, 129908. [CrossRef]
4. Musharavati, F.; Khanmohammadi, S. Performance improvement of a heat recovery system combined with fuel cell and thermoelectric generator: 4E analysis. *Int. J. Hydrogen Energy* **2022**, in press.
5. Cho, Y.H.; Park, J.; Chang, N.; Kim, J. Comparison of Cooling Methods for a Thermoelectric Generator with Forced Convection. *Energies* **2020**, *13*, 3185. [CrossRef]
6. Li, C.; Jiang, F.; Liu, C.; Liu, P.; Xu, J. Present and future thermoelectric materials toward wearable energy harvesting. *Appl. Mater. Today* **2019**, *15*, 543–557. [CrossRef]
7. Kubenova, M.M.; Kuterbekov, K.A.; Balapanov, M.K.; Ishembetov, R.K.; Bekmyrza, K.Z. Some thermoelectric phenomena in copper chalcogenides replaced by lithium and sodium alkaline metals. *Nanomaterials* **2021**, *11*, 2238. [CrossRef] [PubMed]

8. Bell, L.E. Cooling, heating, generating power, and recovering waste heat with thermoelectric systems. *Science* **2008**, *321*, 1457–1461. [CrossRef]
9. Yin, L.-C.; Liu, W.-D.; Li, M.; Sun, Q.; Gao, H.; Wang, D.-Z.; Wu, H.; Wang, Y.-F.; Shi, X.-L.; Liu, Q.; et al. High Carrier Mobility and High Figure of Merit in the CuBiSe2 Alloyed GeTe. *Adv. Energy Mater.* **2021**, *11*, 2102913. [CrossRef]
10. Ao, D.-W.; Liu, W.-D.; Chen, Y.-X.; Wei, M.; Jabar, B.; Li, F.; Shi, X.-L.; Zheng, Z.-H.; Liang, G.-X.; Zhang, X.-H.; et al. Novel Thermal Diffusion Temperature Engineering Leading to High Thermoelectric Performance in Bi2Te3-Based Flexible Thin-Films. *Adv. Sci.* **2022**, *9*, 2103547. [CrossRef]
11. Li, X.; Xie, C.; Quan, S.; Shi, Y.; Tang, Z. Optimization of thermoelectric modules' number and distribution pattern in an automotive exhaust thermoelectric generator. *IEEE Access* **2019**, *7*, 72143–72157. [CrossRef]
12. Spriggs, P.; Wang, Q. Computationally Modelling the Use of Nanotechnology to Enhance the Performance of Thermoelectric Materials. *Energies* **2020**, *13*, 5096. [CrossRef]
13. Shen, Z.-G.; Wu, S.-Y.; Xiao, L. Assessment of the performance of annular thermoelectric couples under constant heat flux condition. *Energy Convers. Manag.* **2017**, *150*, 704–713. [CrossRef]
14. Chen, W.H.; Wu, P.H.; Lin, Y.L. Performance optimization of thermoelectric generators designed by multi-objective genetic algorithm. *Appl. Energy* **2018**, *209*, 211–223. [CrossRef]
15. Fan, L.; Zhang, G.; Wang, R.; Jiao, K. A comprehensive and time-efficient model for determination of thermoelectric generator length and cross-section area. *Energy Convers. Manag.* **2016**, *122*, 85–94. [CrossRef]
16. Zhang, M.; Wang, J.; Tian, Y.; Zhou, Y.; Zhang, J.; Xie, H.; Wu, Z.; Li, W.; Wang, Y. Performance comparison of annular and flat-plate thermoelectric generators for cylindrical hot source. *Energy Rep.* **2021**, *7*, 413–420. [CrossRef]
17. Zhu, W.; Weng, Z.; Li, Y.; Zhang, L.; Zhao, B.; Xie, C.; Shi, Y.; Huang, L.; Yan, Y. Theoretical analysis of shape factor on performance of annular thermoelectric generators under different thermal boundary conditions. *Energy* **2022**, *239*, 122285. [CrossRef]
18. Weng, Z.; Liu, F.; Zhu, W.; Li, Y.; Xie, C.; Deng, J.; Huang, L. Performance improvement of variable-angle annular thermoelectric generators considering different boundary conditions. *Appl. Energy* **2022**, *306*, 118005. [CrossRef]
19. Shen, Z.-G.; Liu, X.; Chen, S.; Wu, S.-Y.; Xiao, L.; Chen, Z.-X. Theoretical analysis on a segmented annular thermoelectric generator. *Energy* **2018**, *157*, 297–313. [CrossRef]
20. Shittu, S.; Li, G.; Zhao, X.; Ma, X.; Akhlaghi, Y.G.; Ayodele, E. High performance and thermal stress analysis of a segmented annular thermoelectric generator. *Energy Convers. Manag.* **2019**, *184*, 180–193. [CrossRef]
21. Wen, Z.F.; Sun, Y.; Zhang, A.B.; Wang, B.L.; Wang, J.; Du, J.K. Performance Analysis of a Segmented Annular Thermoelectric Generator. *J. Electron. Mater.* **2020**, *49*, 4830–4842. [CrossRef]
22. Luo, D.; Wang, R.; Yu, W.; Sun, Z.; Meng, X. Modelling and simulation study of a converging thermoelectric generator for engine waste heat recovery. *Appl. Therm. Eng.* **2019**, *153*, 837–847. [CrossRef]
23. Li, Y.; Wang, S.; Zhao, Y.; Lu, C. Experimental study on the influence of porous foam metal filled in the core flow region on the performance of thermoelectric generators. *Appl. Energy* **2017**, *207*, 634–642. [CrossRef]
24. Yang, W.; Zhu, W.; Li, Y.; Zhang, L.; Zhao, B.; Xie, C.; Yan, Y.; Huang, L. Annular thermoelectric generator performance optimization analysis based on concentric annular heat exchanger. *Energy* **2022**, *239*, 122127. [CrossRef]
25. Zhu, W.; Deng, Y.; Wang, Y.; Shen, S.; Gulfam, R. High-performance photovoltaic-thermoelectric hybrid power generation system with optimized thermal management. *Energy* **2016**, *100*, 91–101. [CrossRef]
26. Li, B.; Huang, K.; Yan, Y.; Li, Y.; Twaha, S.; Zhu, J. Heat transfer enhancement of a modularised thermoelectric power generator for passenger vehicles. *Appl. Energy* **2017**, *205*, 868–879. [CrossRef]
27. Jaworski, M.; Bednarczyk, M.; Czachor, M. Experimental investigation of thermoelectric generator (TEG) with PCM module. *Appl. Therm. Eng.* **2016**, *96*, 527–533. [CrossRef]
28. Sajid, M.; Hassan, I.; Rahman, A. An overview of cooling of thermoelectric devices. *Renew. Sust. Energ. Rev.* **2017**, *78*, 15–22. [CrossRef]
29. Meng, J.-H.; Wu, H.-C.; Wang, T.-H. Optimization of Two-Stage Combined Thermoelectric Devices by a Three-Dimensional Multi-Physics Model and Multi-Objective Genetic Algorithm. *Energies* **2019**, *12*, 2832. [CrossRef]
30. Zhao, Y.; Fan, Y.; Ge, M.; Xie, L.; Li, Z.; Yan, X.; Wang, S. Thermoelectric performance of an exhaust waste heat recovery system based on intermediate fluid under different cooling methods. *Case Stud. Therm. Eng.* **2021**, *23*, 100811. [CrossRef]
31. Luo, D.; Wang, R.; Yu, W.; Zhou, W. A numerical study on the performance of a converging thermoelectric generator system used for waste heat recovery. *Appl. Energy* **2020**, *270*, 115181. [CrossRef]
32. He, W.; Wang, S.; Zhang, X.; Li, Y.; Lu, C. Optimization design method of thermoelectric generator based on exhaust gas parameters for recovery of engine waste heat. *Energy* **2015**, *91*, 1–9. [CrossRef]
33. He, W.; Wang, S.; Lu, C.; Zhang, X.; Li, Y. Influence of different cooling methods on thermoelectric performance of an engine exhaust gas waste heat recovery system. *Appl. Energy* **2016**, *162*, 1251–1258. [CrossRef]
34. Luo, D.; Sun, Z.; Wang, R. Performance investigation of a thermoelectric generator system applied in automobile exhaust waste heat recovery. *Energy* **2022**, *238*, 121816. [CrossRef]
35. He, W.; Guo, R.; Liu, S.; Zhu, K.; Wang, S. Temperature gradient characteristics and effect on optimal thermoelectric performance in exhaust power-generation systems. *Appl. Energy* **2020**, *261*, 114366. [CrossRef]

36. He, W.; Wang, S.; Li, Y.; Zhao, Y. Structural size optimization on an exhaust exchanger based on the fluid heat transfer and flow resistance characteristics applied to an automotive thermoelectric generator. *Energy Convers. Manage.* **2016**, *129*, 240–249. [CrossRef]
37. Yang, Y.; Wang, S.; Zhu, Y. Evaluation method for assessing heat transfer enhancement effect on performance improvement of thermoelectric generator systems. *Appl. Energy* **2020**, *263*, 114688. [CrossRef]
38. Niu, X.; Yu, J.; Wang, S. Experimental study on low-temperature waste heat thermoelectric generator. *J. Power Sources* **2009**, *188*, 621–626. [CrossRef]
39. Ge, M.; Wang, X.; Zhao, Y. Performance analysis of vaporizer tube with thermoelectric generator applied to cold energy recovery of liquefied natural gas. *Energy Convers. Manage.* **2019**, *200*, 112112. [CrossRef]
40. Zhou, Z.-G.; Zhu, D.-S.; Wu, H.-X.; Zhang, H.-S. Modeling, experimental study on the heat transfer characteristics of thermoelectric generator. *J. Therm. Sci.* **2013**, *22*, 48–54. [CrossRef]

Article

Nanostructured Thermoelectric PbTe Thin Films with Ag Addition Deposited by Femtosecond Pulsed Laser Ablation

Alessandro Bellucci [1,*], Stefano Orlando [2], Luca Medici [3], Antonio Lettino [3], Alessio Mezzi [4], Saulius Kaciulis [4] and Daniele Maria Trucchi [1]

[1] Istituto di Struttura della Materia (ISM)—Sez. Montelibretti, DiaTHEMA Laboratory, Consiglio Nazionale delle Ricerche, Via Salaria km 29.300, 00015 Monterotondo, Italy; danielemaria.trucchi@cnr.it
[2] Istituto di Struttura della Materia (ISM)—Sez. Tito Scalo, FemtoLab, Consiglio Nazionale delle Ricerche, Zona Industriale, 85050 Tito, Italy; stefano.orlando@cnr.it
[3] Istituto di Metodologie per l'Analisi Ambientale (IMAA), Consiglio Nazionale delle Ricerche, Zona Industriale, 85050 Tito, Italy
[4] Istituto per lo Studio dei Materiali Nanostrutturati (ISMN)—Sez. Montelibretti, Consiglio Nazionale delle Ricerche, Via Salaria km 29.300, 00015 Monterotondo, Italy; alessio.mezzi@cnr.it (A.M.)
* Correspondence: alessandro.bellucci@ism.cnr.it

Citation: Bellucci, A.; Orlando, S.; Medici, L.; Lettino, A.; Mezzi, A.; Kaciulis, S.; Trucchi, D.M. Nanostructured Thermoelectric PbTe Thin Films with Ag Addition Deposited by Femtosecond Pulsed Laser Ablation. *Energies* **2023**, *16*, 3216. https://doi.org/10.3390/en16073216

Academic Editor: Mahmoud Bourouis

Received: 28 February 2023
Revised: 28 March 2023
Accepted: 30 March 2023
Published: 3 April 2023

Copyright: © 2023 by the authors. Licensee MDPI, Basel, Switzerland. This article is an open access article distributed under the terms and conditions of the Creative Commons Attribution (CC BY) license (https://creativecommons.org/licenses/by/4.0/).

Abstract: Pulsed laser deposition operated by an ultra-short laser beam was used to grow in a vacuum and at room temperature natively nanostructured thin films of lead telluride (PbTe) for thermoelectric applications. Different percentages of silver (Ag), from 0.5 to 20% of nominal concentration, were added to PbTe deposited on polished technical alumina substrates using a multi-target system. The surface morphology and chemical composition were analyzed by Scanning Electron Microscope and X-ray Photoelectron Spectroscopy, whereas the structural characteristics were investigated by X-ray Diffraction. Electrical resistivity as a function of the sample temperature was measured by the four-point probe method by highlighting a typical semiconducting behavior, apart from the sample with the maximum Ag concentration acting as a degenerate semiconductor, whereas the Seebeck coefficient measurements indicate n-type doping for all the samples. The power factor values (up to 14.9 $\mu W\ cm^{-1}\ K^{-2}$ at 540 K for the nominal 10% Ag concentration sample) are competitive for low-power applications on flexible substrates, also presuming the achievement of a large reduction in the thermal conductivity thanks to the native nanostructuring.

Keywords: thermoelectric properties; nanostructuring; pulsed fs-laser deposition; lead telluride

1. Introduction

In the last decade, a great interest has been growing to find a way of developing efficient thermoelectric devices for the conversion of heat into electricity by exploiting the Seebeck effect [1,2], with the aim of re-using the large heat that is typically wasted in several industrial processes. Additionally, thermoelectric generators have been used to produce renewable energy, particularly for the conversion of concentrated solar radiation, both directly [3] or combined with thermionic generators [4,5]. The conversion efficiency of thermoelectric materials is characterized by the dimensionless figure of merit $zT = S^2 \sigma T / \kappa$, where S, σ, T, and κ are the Seebeck coefficient, the electrical conductivity, the absolute temperature, and the thermal conductivity, respectively. Extensive and detailed studies have established, over a long period of time, that to achieve an optimal value of the coefficient zT, it is advisable to follow the strategy of the "phonon glass electron crystal" approach [6], which implies the creation of a material with a tailored structure that has, at the same time, low thermal conductivity (phonon glass), but high electrical conductivity (electron crystal). Consequently, several studies came to find a way to optimize thermoelectric material by maximizing its performance. Nanostructuring is the most diffused approach to improve zT, by enhancing the Seebeck coefficient via the quantum confinement effect and, at the same time, by decreasing thermal conductivity through enhanced phonon scattering at grain boundaries and interfaces [7–10]. Several complex nanostructures have been designed for thermoelectric materials: two-dimensional superlattice structures, one-dimensional nano-systems, and zero-dimensional nanoplates and nanoparticles [11–15].

Compared with bulky systems, the production of thin films has the noteworthy advantage of being compatible with semiconductor technology. Additionally, thin films can be easily applied in the fabrication of thermoelectric nano-devices, the research of which is becoming extremely important thanks to the continuous requirements of miniaturization and flexibility for the novel application scenario of the Internet of Things [16–18]. However, the presently low performance of thermoelectric films in terms of conversion efficiency and temperature stability restricts their real use in the required applications.

Table 1 shows some of the most recent and relevant results obtained from the preparation and characterization of thermoelectric thin films operating at temperatures higher than 350 K. Different approaches and solutions in terms of both materials and techniques were attempted, with the attaining of maximum power factor values in the range of 0.2–4.6 mW m^{-1} K^{-2}, which are typically lower than the values reported for the corresponding bulk material.

Table 1. List of the most recent published works related to the production of thermoelectric nanostructured thin films, displaying material and dopants, technique, power factor at the temperature T, which maximizes the value, and year of publication.

Material/Dopants	Production Technique	Maximum Power Factor (W m^{-1} K^{-2}) at T	Publication Year (Reference)
ZnSb	RF Magnetron Sputtering	2.35×10^{-3} at 533 K	2014 [19]
Bi$_{0.5}$Sb$_{1.5}$Te$_3$	KrF excimer PLD	3.2×10^{-3} at 390 K	2015 [20]
PbTe	Thermal evaporation	$\sim 0.45 \times 10^{-3}$ at 400 K	2015 [21]
ZnSb/Cr	ArF excimer PLD	0.2×10^{-3} at 600 K	2017 [22]
SnTe	Thermal evaporation	1.98×10^{-3} at 823 K	2018 [23]
Cu$_2$Se	XeCl excimer PLD	$\sim 0.2 \times 10^{-3}$ at 580 K	2022 [24]
CoSb$_3$/Ag, Ti	Magnetron Sputtering	0.31×10^{-3} at 623 K	2023 [25]
Sb$_2$Te$_3$/Ag	RF Magnetron Sputtering	4.6×10^{-3} at 373 K	2023 [26]
Sm$_y$(Fe$_x$Ni$_{1-x}$)Sb$_{12}$	Nd: YAG PLD	$\sim 0.38 \times 10^{-3}$ at 525 K	2023 [27]

Among the different thermoelectric materials, lead telluride (PbTe) is a very appealing thermoelectric material for applications in the range of operating temperatures from 300 to 800 K [28]. Several research efforts have been oriented to the improvement of the performance of PbTe bulk alloys through nanostructure optimization [29] and/or the substitution/addition of dopants to adjust both thermal conductivity and carrier concentration [30,31]. However, in the last years, many chalcogenides have lost attraction due to the high cost, the problem of toxicity, and the decreased material availability.

One of the possibilities to be explored to overcome these issues is to use PbTe as thin film, with the main intention to reduce the waste of material and, thus, to decrease the cost of the final device, while maintaining high conversion performance. Many techniques have been developed for the preparation of PbTe thin films [32–34]. Recently, Pulsed Laser Deposition (PLD) has been demonstrated to be a technique capable of fabricating high-quality thermoelectric films [35,36], mainly thanks to its experimental versatility and capability of almost maintaining the stoichiometry in the transfer of mass from the ablated target to the depositing substrate, in a vacuum or in a suitable gas atmosphere.

More specifically, femtosecond (fs) laser ablation in a vacuum has been suggested as a powerful and versatile tool to produce nanoparticles of various materials. With ultra-short pulses in the fs range, the laser intensity on the target is sufficiently high for multi-photon absorption processes to take place; therefore, it becomes possible to fabricate thin films and nanostructures of many materials, along with a low optical absorption coefficient (or large bandgap) at the specific laser wavelength, such as oxides, nitrides, or semiconductors [37,38]. Moreover, the initial laser-heating occurs almost at solid density, leading the matter to extreme temperature and pressure, generating a great number of material states, which cannot be produced using longer pulses of comparable fluence. In addition, laser pulses do not interact with the ejected particles, avoiding unwanted secondary laser–material interactions or reactions.

This study aims to introduce a simple and fast approach to tune the properties of nanostructured thermoelectric materials with the addition of potential dopants for achieving high-performance thin films. The previous works published by our group showed how it is possible using PLD to obtain PbTe-based granular nanocrystalline film, with features at the optimum size, i.e., in the range of 20–100 nm [10,39], for scattering the thermal phonons and, thus, potentially achieving a drastic reduction in thermal conductivity. Despite this promising evidence, the values of the thermoelectric power factor were still not competitive with the performance of the bulk materials. With the scope of finding a strategy for improving the overall performance of the thin films, the specific goal of this work was to investigate the role of Ag incorporation as a possible effective dopant in the PbTe matrix using PLD activated by an ultrashort laser at the femtosecond pulse duration, by evaluating the addition of various percentages of atomic Ag in a multi-layer PbTe/Ag structure.

2. Materials and Methods

The experimental setup of the PLD system consists of an ultra-short Spectra Physics Spitfire Pro XP Ti:Sapphire pulsed laser source (wavelength $\lambda = 800$ nm, pulse duration of ~100 fs; energy of 3.7 ± 0.1 mJ/pulse) in front of a stainless steel vacuum chamber evacuated by a turbo molecular pump (typical background pressure of 5×10^{-7} mbar). The pulse repetition rate was fixed to 100 Hz (the maximum selectable value is 1 kHz). The PbTe and Ag targets (commercial 1-inch diameter targets, 99.999% purity) were placed on a rotating holder during ablation to avoid cratering. The laser beam was focused with an entrance angle of $45°$, by a 1.0 m focal lens. As previously described [35,36], a multi-target system was used to deposit both materials through a PbTe/Ag multi-layered structure. The doping process was properly time-regulated and tuned to obtain a final product with specific Ag atomic percentage composition, promoting the Ag diffusion into the film. For the doping process, a PbTe/Ag multilayer structure was considered an acceptable attempt to reach a heavy doping condition. A doping sequence as $N \times (A(t_1) \cdot B(t_2))$, with A = PbTe, B = Ag, and t_1, t_2, N as variable parameters according to the nominal Ag concentration, was used to perform the deposition. The overall deposition time was 30 min for all the samples. The growth rate was 0.11 μm/min, evaluated by ex-situ measurements of the film thickness. Technical-grade mirror-polished alumina plates were used as deposition substrates. All the substrates were ultrasonically cleaned in n-hexane and positioned on a sample holder maintained at room temperature at ~ 50 mm from the target.

XPS measurements were performed using an ESCALAB 250Xi spectrometer (Thermo Fisher Scientific Ltd., Basingstoke, UK), equipped with a monochromatic Al Kα source (1486.7 eV) for XPS and a six channeltrons detection system. The binding energy (BE) scale was calibrated, positioning the adventitious carbon peak C 1s at BE = 285.0 eV. The photoemission spectra were acquired at constant pass energy of 50 eV. The Avantage v.5.9 software (Thermo Fisher Scientific Ltd., Basingstoke, UK) was used to collect and process the spectra.

Surface morphology was characterized by a Field Emission Scanning Electron Microscopy apparatus (FE-SEM model Supra 40, ZEISS NTS Gmbh, Oberkochen, Germany), to be able to observe the grown materials at a very high resolution, down to nanometric sizes.

Grazing angle X-ray micro-diffraction (μ-XRD) measurements, which are typically used for investigating the structural parameters of surfaces different from the bulk and thin films [40], were performed using a Rigaku D/MAX RAPID (Tokyo, Japan) diffraction system, operating at 40 kV and 30 mA. This instrument is equipped with a Cu Kα source, curved-image plate detector, flat graphite monochromator, variety of beam collimators, and motorized stage (allowing two angular movements, rotation ϕ and revolution ω). The μ-XRD data were collected as two-dimensional images and then converted into Angle (2θ)−Intensity profiles using specific software provided by the producer. The data were acquired using 50 μm collimator, 1 h of collection time, a fixed revolution angle ω (3°), and rotation angles Φ (60°, 75°, 80°, 90°). The peak assignment was made using JCPDS database, whereas the unit-cell parameters of the PbTe samples were refined from the μ-XRD data using the UNITCELL software [41].

The temperature-dependent electrical characterization was performed in a homemade vacuum chamber by using the "four-contact-in-line-points probe" method. This method allows the measurement of sheet resistance under dark conditions and film resistivity when the thickness is known. Sheet resistance is commonly used to characterize thin film materials, with almost uniform thickness. The measurements were performed in a homemade system varying the temperature in the range of 300–600 K under controlled vacuum conditions ($\sim 10^{-3}$ mbar). The value of the film thickness used for the calculation of the electrical resistivity was 3.30 ± 0.05 μm, measured by a commercial profilometer and found almost constant for all the samples. The setup scheme and the details of the method are reported in a previous paper [42].

The Seebeck coefficient measurements were performed using commercial Seebeck Instruments K-20 and SB-100 (MMR technologies, San Jose, CA, USA) with a high impedance amplifier (30 gain) in the temperature range 300–600 K, following the method reported by Ko and Murray [43].

3. Results

The experiments reported in this article follow the work we have undertaken and published on the deposition of thermoelectric PbTe-based thin films via ns- and fs-PLD [35,36]. Our previous experiments verified some fundamental aspects concerning both the PLD methods with related advantages/limitations and some important properties of the materials themselves, which are usually prepared and studied in the form of bulk solid materials.

In particular, an ultrashort laser for PLD is preferred as an activating source if compared to more traditional ns-pulse laser for the following reasons: (i) the rate of ablation and subsequent growth of the film is much higher, and layers of micrometers can be obtained in few minutes; (ii) the material deposited is nanostructured, therefore perfectly functional to request a specific nano-sized structure; (iii) the stoichiometry of the target is maintained without any chemical disproportion; (iv) the material obtained resulted in being crystalline even by keeping the substrate at room temperature; and (v) the adopted multi-target and multi-layered deposition, which is time-regulated, can potentially give the required atomic composition, through the entire thickness of the film.

In the present experiments, we have grown room-temperature PbTe films with the addition of Ag by PLD, operated by an ultra-short fs laser beam, comparing the typical physical–chemical parameters (microstructural, stoichiometric, and electrical) of films at different nominal dopant concentrations, whose expected values were calculated as atomic percentages.

First, the surface chemical composition of the different samples was investigated by XPS analysis. The obtained results evidenced that PbTe films doped with different amounts of Ag were deposited. This result is confirmed by the peak-fitting analysis performed on the spectra shown in Figure 1, shown for PbTe10, as an example. The peaks of Te$3d_{5/2}$ and Pb $4f_{7/2}$ positioned at BE = 137.5 eV and 572.2 eV, respectively, were assigned to PbTe. However, upon contact with air, a thin layer of oxides (few nm) is formed, as evidenced in the spectra deconvolution, which is then promptly removed after a few seconds of ion sputtering (Ar$^+$—1 keV). Instead, the Ag spectrum was characterized by a single Ag$3d_{5/2}$ peak positioned at 368.3 eV, BE value characteristic for Ag^0, thus excluding the formation of secondary phases, such as Ag$_2$Te [21].

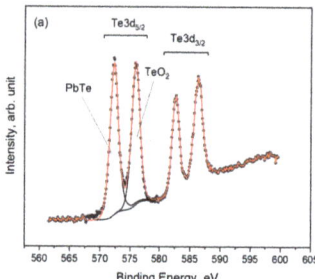 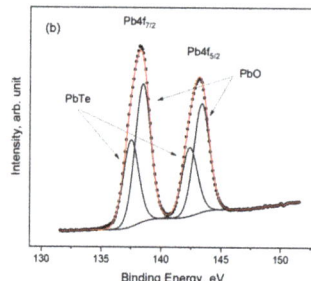 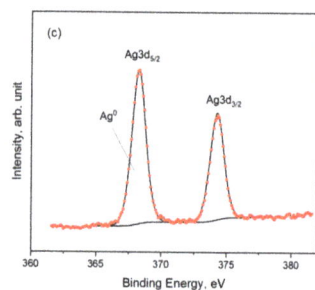

Figure 1. Te3d (**a**), Pb4f (**b**), and Ag3d (**c**) spectra acquired by the investigation of the sample PbTe10.

Table 2 shows the main chemical information derived from the XPS investigation.

Table 2. Values of the XPS atomic concentration of Ag and Pb/Te atomic ratio for the investigated samples.

Sample Name	Nominal [Ag] (%)	[Ag] (at.%) [1]	[Pb]/[Te] (at.%/at.%) [1]
PbTe05	0.5	1.1	1.2
PbTe1	1	2.5	1.2
PbTe2	2	4.6	1.1
PbTe5	5	8.8	0.9
PbTe10	10	14.4	0.7
PbTe20	20	24.9	0.4

[1] Values measured after the sputtering of the first layers to remove contaminants.

From the analysis of the XPS data, it is possible to state that the atomic [Ag] concentration is quite high with respect to the nominal one. This difference is higher when the nominal concentration is low because of the time-dependence of the multi-layer process: for low concentration, the time for a single step is lower and the error increases. However, a linear dependence between the [Pb]/[Te] atomic concentration ratio and the Ag atomic concentration was found (Figure 2a). The amount of Ag corresponding to a stoichiometry close to the correct one ([Pb]/[Te] = 1) is up to 8.8 at.%, whereas the ratio [Pb]/[Te] drastically decreases for [Ag] concentration >8.8 at.% when [Te] is much higher than [Pb]. The large difference from the correct stoichiometry with an excess of Te atoms is surely ascribable to the high number of Ag atoms, which can physically substitute Pb atoms in the Pb–Te system but can also produce doping in interstitial sites or by metallic aggregates. Figure 2b shows the depth profile of the sample PbTe05, which points out how the concentrations of Pb, Te and Ag are quite constant along the thickness of the film apart from the first surface layers (due to carbon and oxygen contaminations).

An evident and characteristic granular nanosized structure is clearly visible in micrographs obtained by SEM, as shown in Figure 3 for samples PbTe05 and PbTe20. The other samples show very similar morphologies and, for brevity, are not reported. An estimation of the size of the granules was carried out via open-source software for imaging analysis (ImageJ, National Institutes of Health, Bethesda, MD, USA) [44]. The size of the granules is randomly distributed from about 220 nm (constituted by few crystallites) down to approximately 30 nm (consisting of single crystallites).

In Figure 4, the XRD spectra of the undoped PbTe (altaite) and that of the sample PbTe10 are shown. We state that the films at different Ag compositions present the same features, without any significant variations in the position and intensity of the reflections reported for the sample PbTe10.

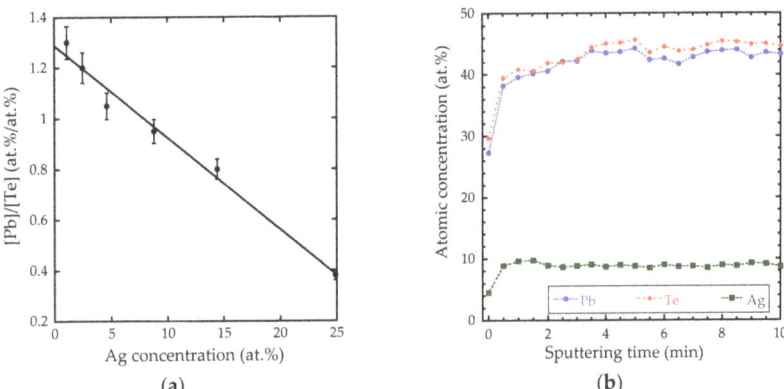

Figure 2. (**a**) [Pb]/[Te] atomic concentration ratio as a function of the Ag concentration; (**b**) XPS depth profile for the sample PbTe05.

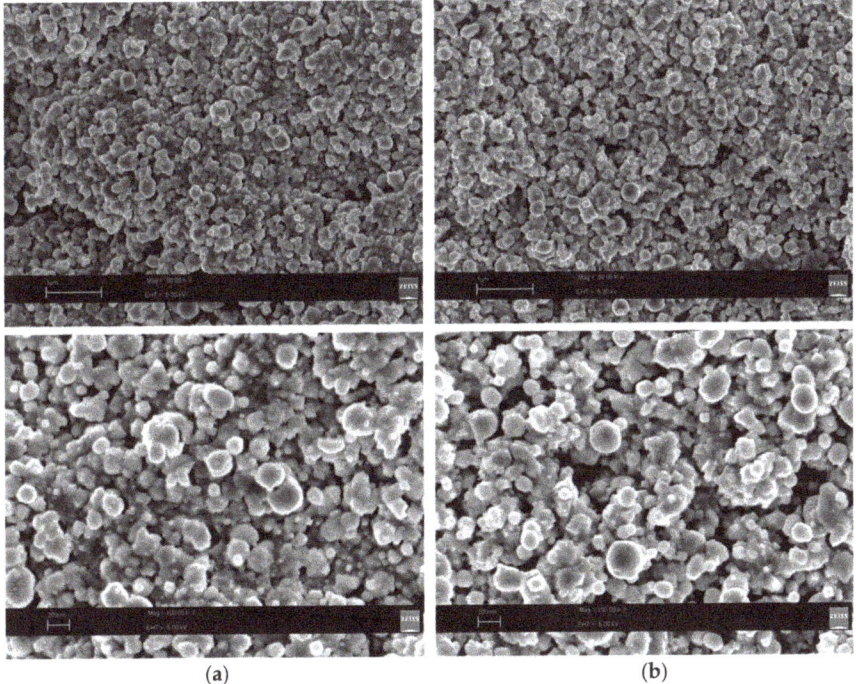

Figure 3. FE-SEM secondary electrons images of thin films: (**a**) PbTe05 and (**b**) PbTe20 at different magnifications. The nanostructured morphology of PbTe grains (~30 to 220 nm) is evident.

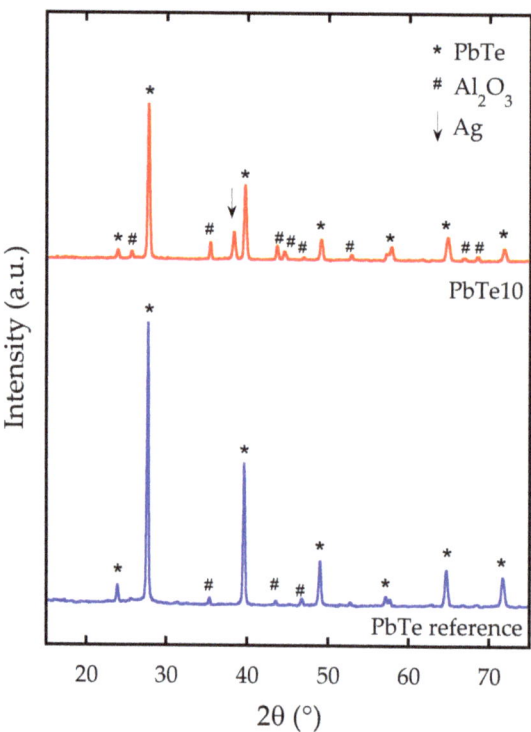

Figure 4. μ-XRD patterns taken at grazing incidence angle of the undoped PbTe, used as a reference, and the sample PbTe10.

It is possible to state that the deposition at room temperature induces the formation of a poly crystalline PbTe film, characterized by the position of the main peaks extremely coherent with the altaite crystal shown in the undoped sample. The crystallite orientation is predominantly in the (200) plane of the cubic phase. All the samples with the Ag addition have a reflection at $2\theta = 38.12°$; this diffraction signal can be attributed to the reflection (111) of Ag nanoparticles [45]. Scherrer's method [46] was applied to the full width at half maximum (FWHM) of the main peak (200) of the cubic PbTe to estimate the size of the crystalline grains. The values of grain size are shown in Table 3 and range from 25 to 39 nm, with a slight increase as a function of the Ag concentration. More interestingly, the extrapolation of the values of the lattice parameter evidences that there is a tendency for reduction when the Ag concentration increases. The decrease in the lattice constant is usually related to the formation of ternary compounds, such as $Pb_{1-x}Ag_xTe$, with the Ag ionic radius being smaller than that of Pb [47]. However, since the lattice parameter does not follow an expected clear trend as a function of the Ag concentration (i.e., Vegard's law), it is most probable that the presence of such compounds is only partially distributed, as reported elsewhere [31]. One of the possible explanations for such behavior is the low solubility limit of Ag for PbTe crystals [48].

From these results, obtained by means of a detailed μ-XRD analysis with a low grazing angle specific for thin film characterization, we can thus conclude that the films are characterized by a granular type of crystalline PbTe nanostructure, with the added Ag playing the role of dopant atoms in the PbTe lattice and creating some metallic dislocations among the grains. Most importantly, the specific particle sizes (25–39 nm) obtained are well befitting to fulfill the specific request for a good thermoelectric material of an effective scattering effect for thermal phonons, as requested to achieve a significant decrease in thermal conductivity inside the lattice of the material examined [7,10].

Table 3. Values of lattice parameter and grain size derived from XRD data for each sample. The parentheses number represents the measurement accuracy.

Sample Name	Lattice Parameter (Å)	Grain Size (nm)
PbTe05	6.454 (3)	25
PbTe1	6.454 (4)	25
PbTe2	6.453 (3)	29
PbTe5	6.449 (3)	39
PbTe10	6.447 (3)	39
PbTe20	6.447 (3)	39

An analysis of the thermoelectric behavior was performed as a function of temperature in the range 300–600 K. Figure 5 shows the Seebeck coefficient (S) for the investigated samples, obtained from ten measurements performed consecutively after repeated heating/cooling cycles to also test the system repeatability (represented by the error bars in the figure). We found for all the samples a negative sign of S, indicating an n-type (electrons) conduction for the majority carriers. At the same time, the absolute value decreases as a function of the Ag concentration, with the sample PbTe20 showing a very low value of S, about one order of magnitude lower than PbTe05. This trend can be expected since the presence of the dopant should increase the carrier concentration. Indeed, according to the electron transport models that are typically used, i.e., the single parabolic band model, S depends on the carrier concentration, such that the lower the carrier concentration, the larger the Seebeck coefficient [7].

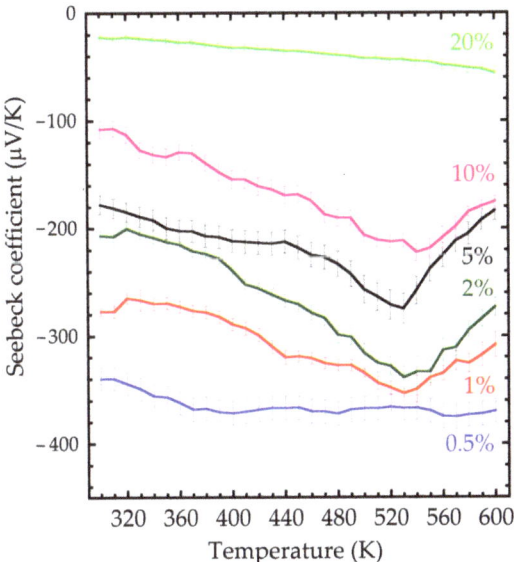

Figure 5. Seebeck coefficient values as a function of the temperature.

Conversely, the Seebeck coefficient shows a particular temperature dependence for all the samples, apart from PbTe05 and PbTe20, with a minimum value between 530 and 550 K. This behavior can be explained with a so-called dynamic doping process, which can occur due to the increase in the carrier concentration when the temperature increases [49]. This performance was fully demonstrated for the addition of Ag atoms in n-type PbTe, where silver compensates the Pb vacancies below 500 K and provides electrons in excess, occupying interstitial sites above 500 K, and thus increasing carrier concentration and/or mobilities [50].

Conversely, the higher nominal Ag concentration induces a lower electrical resistivity (ρ), as shown in Figure 6. The resistivity decreases as a function of the temperature, thus indicating a semiconducting behavior for all the samples apart from the sample PbTe20. The latter shows the electrical resistivity, which increases when the temperature increases, thus pointing out a behavior of a degenerate semiconductor assumed when the Ag concentration is higher than 20 at.%. Conversely to the Seebeck coefficient, we did not find any temperature-dependent change in the electrical resistivity, thus pointing out that future investigations must be conducted to highlight the complex mechanism of the doping needed to completely improve the electronic transport.

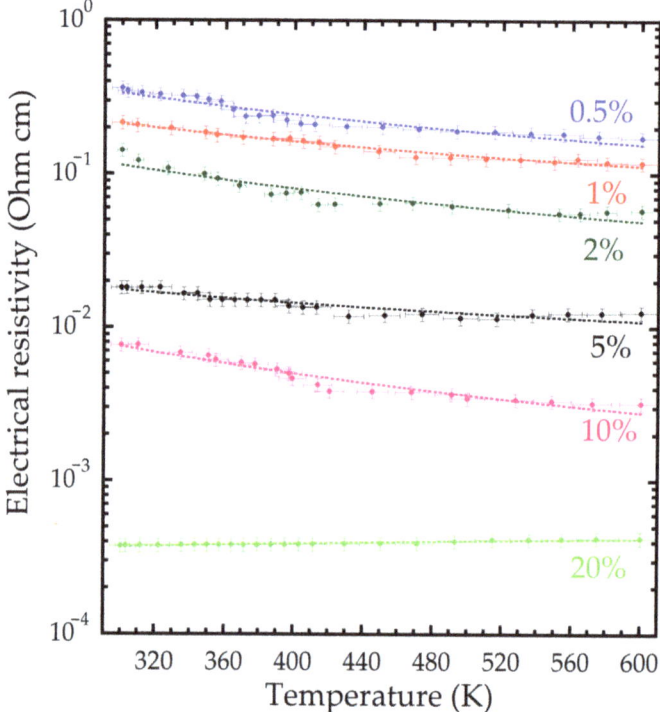

Figure 6. Electrical resistivity values as a function of the temperature.

By combining the electrical conductivity $\sigma = 1/\rho$ and the Seebeck coefficient measurements, it was finally possible to evaluate the material power factor, defined as PF = $S^2 \times \sigma$ (Figure 7).

The PF increases as a function of the dopant concentration up to the nominal concentration of 10%, whereas the trend is opposite for the most doped sample PbTe20 due to the very low value of S. Note that the PF is larger by more than one order of magnitude for the sample PbT10 with respect to the PbTe05 one. It is interesting to state that, apart from PbTe05 and PbTe20, the samples show a maximum for PV ranging from 510 and 540 K, corresponding to the minimum observed for S, thus indicating a specific window for the upper temperature of the operations of a possible device. The highest PF value (14.9 μW K^{-2} cm^{-1} at 540 K) is more than four times the best performance of the PbTe films grown using ns-PLD by our group [35], thanks to a more efficient process of doping performed by the fs-PLD system. Generally, the results achieved in this study are in line with the PF values reported for thermoelectric films produced by magnetron-sputtering deposition (ranging from 0.33 to 23 µW K^{-2} cm^{-1}) [51], and not far from complex nanocomposite PbTe-based bulk systems (up to 30 µW K^{-2} cm^{-1}) [52].

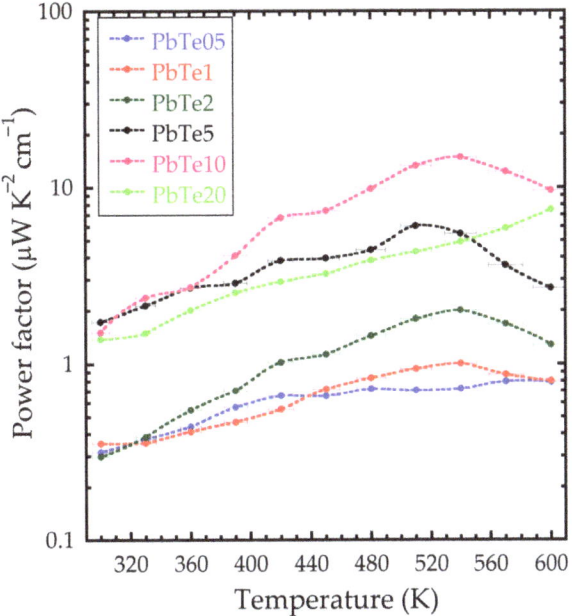

Figure 7. Power factor values as a function of the temperature.

The optimization of the power factor mainly depends on the best trade-off conditions between electrical conductivity and the Seebeck coefficient, which can be obtained with values of the carrier concentration (around 10^{20} cm^{-3}) that are typical of highly doped semiconductors [7]. In our case in which Ag incorporation is applied, the best condition is achieved for a dopant concentration of [Ag] = 14.4 at.%. According to the results of this study, further optimization is difficult to obtain due to the excessive reduction in S if Ag concentration increases, regardless of the improvement in the electrical conductivity. However, co-doping with different elements or even with other complex alloys could be a strategy to be pursued to obtain thin films with improved performance.

To conclude the experimental activity, thermal annealing was performed to evaluate the stability of the thin films, which were left at 600 K for 72 h under vacuum conditions in the same setup used for the electrical resistivity measurements. Any significant change in the structural part of the film was observed, noting that even the electrical parameters did not show significant changes (<5% in terms of absolute value) following the thermal treatment. Therefore, it is possible to state that excellent nanostructured and nano-crystalline films are obtained even with easier depositions at RT. The overall system results are stable from a thermodynamic point of view (the grain structure remains unaltered); although, at present, we have no specific data about any change in the density of the films.

4. Discussion

The characterization of the Ag-added PbTe samples prepared in this work showed that nanostructured materials can be natively deposited by fs-PLD, with an average grain size significantly lower than 100 nm. The process of nanostructuring at dimensions smaller than the mean free path of phonons, while larger than those of charge carriers, allows us to obtain a material with good thermoelectric characteristics [6]. In these conditions, in fact, the phonons are scattered, reducing the lattice's thermal conductivity, while the mobility of charge carriers and electron conduction can remain unchanged.

Therefore, fs-PLD is found to be a powerful technique for producing efficient thermoelectric thin films, combining native nanostructuring with the capability of growing films at RT and at a high deposition rate, as well as being able to dope the materials with the addition of different elements to the starting material structure.

To have an idea of the potential of the developed thin films, a possible estimation of zT can be carried out for the best-produced sample (PbTe10). A zT = 0.64 at the optimum temperature of 540 K is roughly estimated, considering $\kappa \sim 1.25 \times 10^{-2}$ W cm^{-1} K^{-1} as an indicative value of thermal conductivity. This value of κ has been recently reported for a bulk PbTe:Ag system [50]. Furthermore, ultralow values of lattice thermal conductivity (κ_L) have been measured for specific PbTe-AgBiTe$_2$ bulk alloys, equal to ~4.0×10^{-3} W cm^{-1} K^{-1}, which is close to the glass theoretical limit for PbTe [53]. If we suppose achieving a similar reduction in κ_L and, taking into account the calculated electronic thermal conductivity $\kappa_e = L \times \sigma \times T$ of 3.9×10^{-3} at T = 540 K for PbTe10 sample (by fixing L = 2.4×10^{-8} V^2 K^{-2} [31]), a promising value of zT of 1.02 could be obtained for PbTe10.

5. Conclusions

The thermoelectric properties of PbTe thin films, deposited with various Ag concentrations by fs-PLD, were investigated to verify the effectiveness of Ag incorporation to enhance the power factor.

From a correlation between XPS and XRD analyses, we can conclude that the thin films are composed of nanostructured (25–39 nm) crystalline grains of PbTe, in which Ag atoms behave partially as effective dopant, by filling existing vacancies or substituting Pb atoms, and also create metallic nano-agglomerates along the grain dislocations.

Electrical resistivity measurements show the typical behavior of semiconductors, apart from the sample with the highest Ag concentration (24.9 at.%), which displays the behavior of a degenerate semiconductor, increasing the resistivity as a function of the temperature. As was predicted, the resistivity decreases significantly with the increasing atomic percentage of Ag. Regarding the Seebeck coefficient, the measurements point out that electrons are the majority carriers and the absolute value decreases as a function of the Ag concentration, thus indicating an increase in the carrier concentration when the quantity of Ag atoms increases. Interestingly, dynamic doping for the Ag-added PbTe system was identified, leading to a maximization of the performance in the range 510–540 K. Moreover, the material deposited at RT proved to be totally stable, both from a structural, morphological, and electrical point of view, after a thermal treatment for 72 h under the maximum investigated temperature (600 K), showing that this material can grow with excellent structure and properties even under experimental conditions that are very mild and accessible.

A maximum PF value of 14.9 µW K^{-2} cm^{-1} was recorded at 540 K for the sample with an Ag concentration of 14.4 at.%, improving the performance of the PbTe films obtained previously with deposition via both ns- and fs-PLD by our group. However, these values are still lower than the best results obtained by other approaches (in terms of both materials and techniques) and, above all, by bulky solutions. Therefore, other strategies, including co-doping with different elements or even complex alloys, should be considered for achieving a further improvement of the thermoelectric properties.

Finally, we can conclude that the deposition method of fs-PLD has proven to be extremely advantageous, both as regards the high growth rate (i.e., 110 nm/min), both for obtaining crystalline material of cubic PbTe with silver incorporation, even at room temperature, and for the unique characteristic of giving rise to nano-sized granular structures in the range 25–200 nm, as demonstrated by detailed structural and morphological investigations. In the future, measurements of thermal conductivity will be performed to measure the effective zT of the produced thin films.

Author Contributions: A.B.: conceptualization, methodology, investigation, data curation, visualization, writing—original draft preparation; S.O.: methodology; L.M., A.L., A.M. and S.K.: investigation; D.M.T.: conceptualization, validation, writing—review and editing, supervision. All authors have read and agreed to the published version of the manuscript.

Funding: This research received no external funding.

Data Availability Statement: Not applicable.

Acknowledgments: The authors thank Emilia Cappelli for her initial contribution to this work.

Conflicts of Interest: The authors declare no conflict of interest.

References

1. Champier, D. Thermoelectric generators: A review of applications. *Energy Convers. Manag.* **2017**, *140*, 167–181. [CrossRef]
2. Tritt, T.M.; Subramanian, M.A. Thermoelectric Materials, Phenomena, and Applications: A Bird's Eye View. *MRS Bull.* **2006**, *31*, 188–198. [CrossRef]
3. Baranowski, L.L.; Snyder, G.J.; Toberer, E.S. Concentrated solar thermoelectric generators. *Energy Environ. Sci.* **2012**, *5*, 9055. [CrossRef]
4. Trucchi, D.M.; Bellucci, A.; Girolami, M.; Calvani, P.; Cappelli, E.; Orlando, S.; Polini, R.; Silvestroni, L.; Sciti, D.; Kribus, A. Solar Thermionic-Thermoelectric Generator (ST2G): Concept, Materials Engineering, and Prototype Demonstration. *Adv. Energy Mater.* **2018**, *8*, 1802310. [CrossRef]
5. Bellucci, A.; Girolami, M.; Mastellone, M.; Serpente, V.; Trucchi, D.M. Upgrade and present limitations of solar thermionic-thermoelectric technology up to 1000 K. *Sol. Energy Mater. Sol. Cells* **2021**, *223*, 110982. [CrossRef]
6. Dresselhaus, M.S.; Chen, G.; Tang, M.Y.; Yang, R.G.; Lee, H.; Wang, D.Z.; Ren, Z.F.; Fleurial, J.P.; Gogna, P. New Directions for Low-Dimensional Thermoelectric Materials. *Adv. Mater.* **2007**, *19*, 1043–1053. [CrossRef]
7. Snyder, G.J.; Toberer, E.S. Complex thermoelectric materials. *Nat. Mater.* **2008**, *7*, 105–114. [CrossRef]
8. Kanatzidis, M.G. Nanostructured Thermoelectrics: The New Paradigm? *Chem. Mater.* **2010**, *22*, 648–659. [CrossRef]
9. Venkatasubramanian, R.; Silvola, E.; Colpitts, T.; O'Quinn, B. Thin-film thermoelectric devices with high room-temperature figures of merit. *Nature* **2001**, *413*, 597–602. [CrossRef]
10. Biswas, K.; He, J.; Blum, I.D.; Wu, C.I.; Hogan, T.P.; Seidman, D.N.; Dravid, V.P.; Kanatzidis, M.G. High-performance bulk thermoelectrics with all-scale hierarchical architectures. *Nature* **2012**, *489*, 414–418. [CrossRef]
11. Yang, L.; Chen, Z.-G.; Han, G.; Hong, M.; Zou, Y.; Zou, J. High-performance thermoelectric Cu_2Se nanoplates through nanostructure engineering. *Nano Energy* **2015**, *16*, 367–374. [CrossRef]
12. Zhang, Y.; Mehta, R.J.; Belley, M.; Han, L.; Ramanath, G.; Borca-Tasciuc, T. Lattice thermal conductivity diminution and high thermoelectric power factor retention in nanoporous macroassemblies of sulfur-doped bismuth telluride nanocrystals. *Appl. Phys. Lett.* **2012**, *100*, 193113. [CrossRef]
13. Chowdhury, I.; Prasher, R.; Lofgreen, K.; Chrysler, G.; Narasimhan, S.; Mahajan, R.; Koester, D.; Alley, R.; Venkatasubramanian, R. On-chip cooling by superlattice-based thin-film thermoelectrics. *Nat. Nanotechnol.* **2009**, *4*, 235–238. [CrossRef]
14. Elyamny, S.; Dimaggio, E.; Magagna, S.; Narducci, D.; Pennelli, G. High Power Thermoelectric Generator Based on Vertical Silicon Nanowires. *Nano Lett.* **2020**, *20*, 4748–4753. [CrossRef]
15. Blackburn, J.L.; Ferguson, A.J.; Cho, C.; Grunlan, J.C. Carbon-Nanotube-Based Thermoelectric Materials and Devices. *Adv. Mater.* **2018**, *30*, 1704386. [CrossRef]
16. Li, X.; Cai, K.; Gao, M.; Du, Y.; Shen, S. Recent advances in flexible thermoelectric films and devices. *Nano Energy* **2021**, *89*, 106309. [CrossRef]
17. Xie, H.; Zhang, Y.; Gao, P. Thermoelectric-Powered Sensors for Internet of Things. *Micromachines* **2022**, *14*, 31. [CrossRef]
18. Han, C.; Tan, G.; Varghese, T.; Kanatzidis, M.G.; Zhang, Y. High-Performance PbTe Thermoelectric Films by Scalable and Low-Cost Printing. *ACS Energy Lett.* **2018**, *3*, 818–822. [CrossRef]
19. Fan, P.; Fan, W.-F.; Zheng, Z.-H.; Zhang, Y.; Luo, J.-T.; Liang, G.-X.; Zhang, D.-P. Thermoelectric properties of zinc antimonide thin film deposited on flexible polyimide substrate by RF magnetron sputtering. *J. Mater. Sci. Mater. Electron.* **2014**, *25*, 5060–5065. [CrossRef]
20. Symeou, E.; Pervolaraki, M.; Mihailescu, C.N.; Athanasopoulos, G.I.; Papageorgiou, C.; Kyratsi, T.; Giapintzakis, J. Thermoelectric properties of Bi0.5Sb1.5Te3 thin films grown by pulsed laser deposition. *Appl. Surf. Sci.* **2015**, *336*, 138–142. [CrossRef]
21. Bala, M.; Gupta, S.; Tripathi, T.S.; Varma, S.; Tripathi, S.K.; Asokan, K.; Avasthi, D.K. Enhancement of thermoelectric power of PbTe: Ag nanocomposite thin films. *RSC Adv.* **2015**, *5*, 25887–25895. [CrossRef]
22. Bellucci, A.; Mastellone, M.; Girolami, M.; Orlando, S.; Medici, L.; Mezzi, A.; Kaciulis, S.; Polini, R.; Trucchi, D.M. ZnSb-based thin films prepared by ns-PLD for thermoelectric applications. *Appl. Surf. Sci.* **2017**, *418*, 589–593. [CrossRef]
23. Xu, S.; Zhu, W.; Zhao, H.; Xu, L.; Sheng, P.; Zhao, G.; Deng, Y. Enhanced thermoelectric performance of SnTe thin film through designing oriented nanopillar structure. *J. Alloys Compd.* **2018**, *737*, 167–173. [CrossRef]
24. Wang, A.; Xue, Y.; Wang, J.; Yang, X.; Wang, J.; Li, Z.; Wang, S. High thermoelectric performance of Cu_2Se-based thin films with adjustable element ratios by pulsed laser deposition. *Mater. Today Energy* **2022**, *24*, 100929. [CrossRef]

25. Wei, M.; Ma, H.L.; Nie, M.Y.; Li, Y.Z.; Zheng, Z.H.; Zhang, X.H.; Fan, P. Enhanced Thermoelectric Performance of CoSb$_3$ Thin Films by Ag and Ti Co-Doping. *Materials* **2023**, *16*, 1271. [CrossRef]
26. Thaowonkaew, S.; Insawang, M.; Vora-ud, A.; Horprathum, M.; Muthitamongkol, P.; Maensiri, S.; Kumar, M.; Phan, T.B.; Seetawan, T. Effect of substrate rotation and rapid thermal annealing on thermoelectric properties of Ag-doped Sb$_2$Te$_3$ thin films. *Vacuum* **2023**, *211*, 111920. [CrossRef]
27. Latronico, G.; Mele, P.; Sekine, C.; Wei, P.S.; Singh, S.; Takeuchi, T.; Bourges, C.; Baba, T.; Mori, T.; Manfrinetti, P.; et al. Effect of the annealing treatment on structural and transport properties of thermoelectric Sm$_y$(Fe$_x$Ni$_{1-x}$)4Sb$_{12}$ thin films. *Nanotechnology* **2023**, *34*, 115705. [CrossRef]
28. Dughaish, Z.H. Lead telluride as a thermoelectric material for thermoelectric power generation. *Phys. B Condens. Matter* **2002**, *322*, 205–223. [CrossRef]
29. Pei, Y.; Wang, H.; Snyder, G.J. Band Engineering of Thermoelectric Materials. *Adv. Mater.* **2012**, *24*, 6125–6135. [CrossRef]
30. Takagiwa, Y.; Pei, Y.; Pomrehn, G.; Snyder, G.J. Dopants effect on the band structure of PbTe thermoelectric material. *Appl. Phys. Lett.* **2012**, *101*, 092102. [CrossRef]
31. Dow, H.S.; Oh, M.W.; Kim, B.S.; Park, S.D.; Min, B.K.; Lee, H.W.; Wee, D.M. Effect of Ag or Sb addition on the thermoelectric properties of PbTe. *J. Appl. Phys.* **2010**, *108*, 113709. [CrossRef]
32. Ito, M.; Seo, W.-S.; Koumoto, K. Thermoelectric properties of PbTe thin films prepared by gas evaporation method. *J. Mater. Res.* **1999**, *14*, 209–212. [CrossRef]
33. Shing, Y.H.; Chang, Y.; Mirshafii, A.; Hayashi, L.; Roberts, S.S.; Josefowicz, J.Y.; Tran, N. Sputtered Bi$_2$Te$_3$ and PbTe thin films. *J. Vac. Sci. Technol. A* **1983**, *1*, 503–506. [CrossRef]
34. Dauscher, A.; Dinescu, M.; Boffoué, O.M.; Jacquot, A.; Lenoir, B. Temperature-dependant growth of PbTe pulsed laser deposited films on various substrates. *Thin Solid Film.* **2006**, *497*, 170–176. [CrossRef]
35. Cappelli, E.; Bellucci, A.; Medici, L.; Mezzi, A.; Kaciulis, S.; Fumagalli, F.; Di Fonzo, F.; Trucchi, D.M. Nano-crystalline Ag–PbTe thermoelectric thin films by a multi-target PLD system. *Appl. Surf. Sci.* **2014**, *336*, 283–289. [CrossRef]
36. Bellucci, A.; Cappelli, E.; Orlando, S.; Medici, L.; Mezzi, A.; Kaciulis, S.; Polini, R.; Trucchi, D.M. Fs-pulsed laser deposition of PbTe and PbTe/Ag thermoelectric thin films. *Appl. Phys. A Mater. Sci. Process.* **2014**, *117*, 401–407. [CrossRef]
37. Amoruso, S.; Ausanio, G.; Bruzzese, R.; Lanotte, L.; Scardi, P.; Vitiello, M.; Wang, X.T. Synthesis of nanocrystal films via femtosecond laser ablation in vacuum. *J. Phys. Condens. Matter* **2006**, *18*, L49–L53. [CrossRef]
38. Bellucci, A.; Orlando, S.; Girolami, M.; Mastellone, M.; Serpente, V.; Paci, B.; Generosi, A.; Mezzi, A.; Kaciulis, S.; Polini, R.; et al. Aluminum (Oxy)nitride thermoelectric thin films grown by fs-PLD as electron emitters for thermionic applications. *AIP Conf. Proc.* **2021**, *2416*, 020004. [CrossRef]
39. Qiu, B.; Bao, H.; Zhang, G.; Wu, Y.; Ruan, X. Molecular dynamics simulations of lattice thermal conductivity and spectral phonon mean free path of PbTe: Bulk and nanostructures. *Comput. Mater. Sci.* **2012**, *53*, 278–285. [CrossRef]
40. Sakata, O.; Nakamura, M. Grazing Incidence X-Ray Diffraction. In *Surface Science Techniques*; Bracco, G., Holst, B., Eds.; Springer: Berlin/Heidelberg, Germany, 2013; pp. 165–190.
41. Holland, T.J.B.; Redfern, S.A.T. UNITCELL: A nonlinear least-squares program for cell-parameter refinement and implementing regression and deletion diagnostics. *J. Appl. Crystallogr.* **1997**, *30*, 84. [CrossRef]
42. Trucchi, D.M.; Zanza, A.; Bellucci, A.; Marotta, V.; Orlando, S. Photoconductive and photovoltaic evaluation of In$_2$O$_3$–SnO$_2$ multilayered thin-films deposited on silicon by reactive pulsed laser ablation. *Thin Solid Film.* **2010**, *518*, 4738–4742. [CrossRef]
43. Ko, D.K.; Murray, C.B. Probing the Fermi Energy Level and the Density of States Distribution in PbTe Nanocrystal (Quantum Dot) Solids by Temperature-Dependent Thermopower Measurements. *ACS Nano* **2011**, *5*, 4810–4817. [CrossRef] [PubMed]
44. Schneider, C.A.; Rasband, W.S.; Eliceiri, K.W. NIH Image to ImageJ: 25 years of image analysis. *Nat. Methods* **2012**, *9*, 671–675. [CrossRef] [PubMed]
45. Swetha, V.; Lavanya, S.; Sabeena, G.; Pushpalaksmi, E.; Jenson, S.J.; Annadurai, G. Synthesis and Characterization of Silver Nanoparticles from *Ashyranthus aspera* Extract for Antimicrobial Activity Studies. *J. Appl. Sci. Environ. Manag.* **2020**, *24*, 1161–1167. [CrossRef]
46. Jagodzinski, H.; Klug, H.P.; Alexander, L.E. X-ray Diffraction Procedures for Polycrystalline and Amorphous Materials, 2. Auflage. John Wiley & Sons, New York-Sydney-Toronto 1974, 966 Seiten, Preis: £18.55. *Ber. Bunsenges. Phys. Chem.* **1975**, *79*, 553.
47. Miotkowska, S.; Dynowska, E.; Miotkowski, I.; Szczerbakow, A.; Witkowska, B.; Kachniarz, J.; Paszkowicz, W. The lattice constants of ternary and quaternary alloys in the PbTe–SnTe–MnTe system. *J. Cryst. Growth* **1999**, *200*, 483–489. [CrossRef]
48. Sharov, M.K. Silver solubility in PbTe crystals. *Inorg. Mater.* **2008**, *44*, 569–571. [CrossRef]
49. Pei, Y.; May, A.F.; Snyder, G.J. Self-Tuning the Carrier Concentration of PbTe/Ag$_2$Te Composites with Excess Ag for High Thermoelectric Performance. *Adv. Energy Mater.* **2011**, *1*, 291–296. [CrossRef]
50. Wang, S.; Chang, C.; Bai, S.; Qin, B.; Zhu, Y.; Zhan, S.; Zheng, J.; Tang, S.; Zhao, L.D. Fine Tuning of Defects Enables High Carrier Mobility and Enhanced Thermoelectric Performance of n-Type PbTe. *Chem. Mater.* **2023**, *35*, 755–763. [CrossRef]
51. Zhang, Z.; Gurtaran, M.; Li, X.; Un, H.I.; Qin, Y.; Dong, H. Characterization of Magnetron Sputtered BiTe-Based Thermoelectric Thin Films. *Nanomaterials* **2023**, *13*, 208. [CrossRef]

52. Rogacheva, E.I.; Krivulkin, I.M.; Nashchekina, O.N.; Sipatov, A.Y.; Volobuev, V.A.; Dresselhaus, M.S. Percolation transition of thermoelectric properties in PbTe thin films. *Appl. Phys. Lett.* **2001**, *78*, 3238–3240. [CrossRef]
53. Zhu, H.; Zhang, B.; Zhao, T.; Zheng, S.; Wang, G.; Wang, G.; Lu, X.; Zhou, X. Achieving glass-like lattice thermal conductivity in PbTe by $AgBiTe_2$ alloying. *Appl. Phys. Lett.* **2022**, *121*, 241903. [CrossRef]

Disclaimer/Publisher's Note: The statements, opinions and data contained in all publications are solely those of the individual author(s) and contributor(s) and not of MDPI and/or the editor(s). MDPI and/or the editor(s) disclaim responsibility for any injury to people or property resulting from any ideas, methods, instructions or products referred to in the content.

Article

Influence of Charge Transfer on Thermoelectric Properties of Endohedral Metallofullerene (EMF) Complexes

Majed Alshammari [1], Turki Alotaibi [1], Moteb Alotaibi [2] and Ali K. Ismael [3,*]

1 Physics Department, College of Science, Jouf University, Sakakah 11942, Saudi Arabia
2 Department of Physics, College of Science and Humanities in Al-Kharj, Prince Sattam bin Abdulaziz University, Al-Kharj 11942, Saudi Arabia
3 Department of Physics, Lancaster University, Lancaster LA1 4YB, UK
* Correspondence: k.ismael@lancaster.ac.uk; Tel.: +44-(0)-1524-593059

Abstract: A considerable potential advantage of manufacturing electric and thermoelectric devices using endohedral metallofullerenes (EMFs) is their ability to accommodate metallic moieties inside their cavities. Published experimental and theoretical works have explained the usefulness of this resilience feature for improving the electrical conductance and thermopower. Through thorough theoretical investigations of three EMF complexes employing three different metallic moieties involving Sc_3C_2, Sc_3N, and Er_3N and their configurations on a gold (111) surface, this research demonstrates that the thermoelectric properties of these molecular complexes can be tuned by taking advantage of the charge transfer from metallic moieties to Ih-C_{80} cages. Mulliken, Hirshfeld, and Voronoi simulations articulate that the charge migrates from metallic moieties to cages; however, the amount of the transferred charge depends on the nature of the moiety within the complex.

Keywords: thermoelectric; power factor; EMFs; charge transfer; EMF complex

Citation: Alshammari, M.; Alotaibi, T.; Alotaibi, M.; Ismael, A.K. Influence of Charge Transfer on Thermoelectric Properties of Endohedral Metallofullerene (EMF) Complexes. *Energies* 2023, *16*, 4342. https://doi.org/10.3390/en16114342

Academic Editor: Diana Enescu

Received: 9 February 2023
Revised: 13 April 2023
Accepted: 28 April 2023
Published: 26 May 2023

Copyright: © 2023 by the authors. Licensee MDPI, Basel, Switzerland. This article is an open access article distributed under the terms and conditions of the Creative Commons Attribution (CC BY) license (https:// creativecommons.org/licenses/by/ 4.0/).

1. Introduction

Charge transfer (CT), electron transfer (ET), and donor–acceptor (DA) complexes have long been the focus of investigation. Consequently, charge transfer perception is essential in many organic devices, because of its various uses in many disciplines involving chemistry, physics, materials science, medicine, and biology. For example, CT has been extensively explored in organic solar cells [1–3], water splitting devices, [4] and single molecule electronics [5–9]. Similarly, there have been varying types of donors and acceptors in complicated charge transfer research [10–13]. The chemical nature of the molecule determines whether a molecule behaves as a donor or acceptor. In such systems, an electron-rich donor commonly acts as the receptor and the acceptor is often electron-deficient. In measurement methods, the charge transfer through molecular systems is classified into two categories: CT in Donor–Bridge–Acceptor (DBA) molecules and CT in Metal–Bridge–Metal (MBM) junctions [14–16].

To probe the charge transfer and density functional theory (DFT), analysis can be employed to determine the nature of two molecular segments (i.e., molecule) in complexes based on their electronic structures, as illustrated in Figure 1. Donor–acceptor interaction energy calculation within the DFT framework plays a crucial role in studying charge transfer behaviours inside a molecular system. DFT analyses have been widely used to investigate CT complexes [17–19].

In the present research, we explore the electronic properties of three donor–acceptor complexes. The major investigation here is dedicated to the analysis of three distinct methodologies including the Mulliken population [20], Hirshfeld [21], and Voronoi [22]. These methods were used to trace down the charge transfer between the molecular segments (see Figure 1). CT calculations were first performed in isolated systems (i.e., gas phase), and then on a Au (111) surface (see Section S3 of the SI).

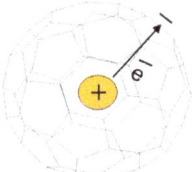

Figure 1. Schematic illustration of donor–acceptor complex Li@C$_{60}$. Li cation is positively charged (donor), while C$_{60}$ cage is negatively charged (acceptor).

Figure 2 below illustrates the anatomy of three complexes, each of which consists of two molecular segments involving a metallic moiety such as Sc$_3$C$_2$, Sc$_3$N, Er$_3$N, and Ih-C$_{80}$ cages. When the metallic moiety is encapsulated inside the fullerene cage, the outcome is endohedral metallofullerene (EMF) complexes. The current research investigates the electronic structure of three EMFs, i.e., Sc$_3$C$_2$@C$_{80}$, Sc$_3$N@C$_{80}$, and Er$_3$N@C$_{80}$; examples of three EMF complexes and an empty fullerene are shown in Supplementary Figures S1 and S2.

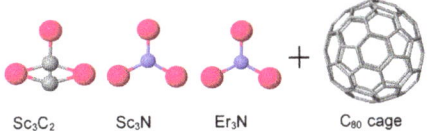

Figure 2. Schematic illustration of three metallic moieties including Sc$_3$C$_2$, Sc$_3$N, Er$_3$N, and Ih-C$_{80}$ cages (Note: All structures are fully optimised). Insertion of the metallic moiety inside the cage yields EMF complexes such as Sc$_3$C$_2$@C$_{80}$, Sc$_3$N@C$_{80}$, and Er$_3$N@C$_{80}$.

2. Computational Methods

All the theoretical simulations were carried out by employing the density functional (DFT) code SIESTA [23]. The optimum geometries of isolated EMFs were obtained by relaxing the molecules until all forces on the atoms were less than 0.01 eV/Å (for more detail, see Supplementary Figures S1 and S2). A double-zeta plus polarization orbital basis set was used, with norm-conserving pseudopotentials and the local density approximation (LDA) exchange with a functional correlation, and to define the real space grid, an energy cut-off of 250 Rydberg was used. Results using GGA were also calculated and found that the resulting functions were comparable [24,25], with results obtained employing LDA functional exchange (see Section 1). To simulate the likely contact configuration during a break-junction experiment, we employed leads constructed from 6 layers of Au (111), each containing 30 gold atoms and further terminated with a pyramid of gold atoms.

To determine the optimum distance of EMF complexes attaching to the Au (111) metals, density functional theory and the counterpoise method were used, which removes basis set superposition errors (BSSEs). The binding distance was defined as the distance between the gold surface and the EMF complex. The ground state energy of the total system was calculated using SIESTA and is denoted E_{AB}^{AB}. The energy of each monomer was then calculated in a fixed basis, which is achieved through the use of ghost atoms in SIESTA (see Section 4). Since the energy of the individual complex in the presence of the fixed basis is defined as E_A^{AB} and for the isolated gold as E_B^{AB}, the binding energy $\Delta(\theta)$, is then calculated using the following equation [26–28]:

$$\text{Binding Energy} = E_{AB}^{AB} - E_A^{AB} - E_B^{AB} \quad (1)$$

3. Results and Discussion

The electronic properties of the three EMF complexes involving Sc$_3$C$_2$@C$_{80}$, Sc$_3$N@C$_{80}$, and Er$_3$N@C$_{80}$ were simulated employing both density functional theory (DFT) and quantum transport theory. To have a deep understanding of thermoelectric properties, the wave function of the investigated complexes, i.e., the lowest unoccupied orbitals (LUMO) and the highest occupied molecular orbitals (HOMO), along with their energies, are explored, as illustrated in Supplementary Figures S3–S5. These isosurface plots clearly demonstrate a significant weight on metallic moieties Sc_3C_2, Sc_3N, and Er_3N, in contrast to Ih-C_{80} cages. The significant weight occurs on the LUMO orbitals and it is well-known that these complexes possess LUMO dominated transport. This denotes that metallic moieties play a pivotal role in tuning the electronic properties of the EMF complexes.

As a first step, we investigated the charge transfer through these EMF complexes. Charge calculations are common practise in chemical science measurements and calculations. We shall first discuss charge transfer analyses in the gas phase for the three EMFs. We evaluate the net charge transfer from metallic moieties Sc_3C_2, Sc_3N, and Er_3N to Ih-C_{80} cages using three different DFT analyses methods Mulliken, Hirshfeld, and Voronoi (see Section S3 in the SI).

Table 1 below demonstrates that the three metallic moieties donate electrons to the C_{80} cage. However, the total number of the transferred electrons depends on the chemical nature (i.e., atom species) and geometrical shape of the metallic moiety. We find that the donation of erbium nitride Er_3N is the highest followed by scandium nitride Sc_3N and then scandium carbide Sc_3C_2. Furthermore, the charge transfer through the three EMF complexes follows the order $Er_3N@C_{80} > Sc_3N@C_{80} > Sc_3C_2@C_{80}$ for Mulliken, Hirshfeld, and Voronoi analyses.

Table 1. Gas phase, charge transfer calculations employing Mulliken, Hirshfeld, and Voronoi methods of $Sc_3C_2@C_{80}$, $Sc_3N@C_{80}$, and $Er_3N@C_{80}$ complexes. The total number of electrons transferred from metallic moieties (with a charge of $+|e|$) to Ih-C_{80} cages (with a charge of $-|e|$) to form complexes. Note: loss–gain differences gained by C_2, N, and N (numbers in brackets) of $Sc_3C_2@C_{80}$, $Sc_3N@C_{80}$, and $Er_3N@C_{80}$ complexes.

Metallic Moiety	Mulliken		Hirshfeld		Voronoi	
	moiety	cage	moiety	cage	moiety	cage
Sc_3C_2	+1.40	−1.14	+1.15	−0.83	+1.06	−0.72
C_2	(−0.26)	-	(−0.32)	-	(−0.34)	-
Sc_3N	+1.50	−1.26	+1.31	−0.98	+1.27	−0.96
N	(−0.24)	-	(−0.33)	-	(−0.31)	-
Er_3N	+6.96	−5.14	+7.48	−6.14	+7.14	−5.82
N	(−1.82)	-	(−1.34)	-	(−1.32)	-

Surprisingly, there is a difference between the total number of the donated and gained electrons through the complexation (i.e., encapsulating the moiety inside cage). For example, the scandium carbide Sc_3C_2 donates 1.40 electrons to the cage; however, only 1.14 is indeed gained by the cage (loss–gain difference), and this occurs through all the EMF complexes. To answer this question, we tracked down the CT from the donor to receptor, atom by atom. The tracking analyses suggest the missing electrons are gained by the metallic moiety itself.

To accommodate this, we find that loss–gain differences are indeed gained by the moieties. For instance, through $Sc_3C_2@C_{80}$ complexation, the two carbon atoms of Sc_3C_2, gain 0.26, 0.32, and 0.34 electrons from the moiety's donation. Similarly, the nitrogen atoms of Sc_3N and Er_3N gain 0.24, 0.33, and 0.31 and 1.82, 1.34, and 1.32 electrons, respectively, when they form complexes with Ih-C_{80} cages (above analyses evaluated via the Mulliken population, Hirshfeld, and Voronoi). It should be noted that the electron travelled from the moiety to the Ih-C_{80} cage has a significant effect on the conductance G and thermopower S; more details have been given previously [20,29]. Moreover, it should be noted that the total number of electrons donated by erbium nitride Er_3N is significantly larger than of Sc_3C_2 and Sc_3N moieties; we will discuss that later.

To mimic the likely metal–organic contact configuration during a scanning tunnelling microscope break-junction measurement (STM-BJ), we now repeat the above analysis for the case when the EMF forms a complex on a gold surface. Calculations in this section are for three parameters: metallic moiety, Au surface (2nd, 4th, and 6th columns), and a Ih-C_{80} cage (3rd, 5th, and 7th columns). Table 2 suggests that the electronic charge travels from both the metallic moiety and the gold surface to the Ih-C_{80} cage for the three configurations (i.e., EMF complex + Au surface). This behaviour is expected to occur as both the moiety and Au are metals, unlike the cage.

Table 2. On a gold surface, charge transfer calculations employing the Mulliken, Hirshfeld, and Voronoi methods of $Sc_3C_2@C_{80}$, $Sc_3N@C_{80}$, and $Er_3N@C_{80}$ complexes. The total number of electrons transferred from metallic moieties and Au surface (with a charge of $+|e|$) to Ih-C_{80} cages (with a charge of $-|e|$) to form complex Au junctions. Note: numbers in brackets correspond to Au donation ($+|e|$) and C_2, N gaining ($-|e|$).

Moiety + Au	Mulliken		Hirshfeld		Voronoi	
	moiety	cage	moiety	cage	moiety	cage
Sc_3C_2	+1.33	−1.39	+0.93	−0.81	+0.97	−0.79
Au, C_2	(+0.3, −0.24)	-	(+0.2, −0.32)	-	(+0.18, −0.36)	-
Sc_3N	+2.15	−2.09	+1.07	−0.98	+1.04	−1.02
Au, N	(+0.22, −0.28)	-	(+0.23, −0.32)	-	(+0.24, −0.26)	-
Er_3N	+6.53	−5.20	+6.96	−5.80	+6.66	−5.60
Au, N	(+0.24, −1.57)	-	(+0.28, −1.44)	-	(+0.29, −1.35)	-

Table 2 explains that both the Sc_3C_2 and Au lose (+) electrons, and in total, their donation is 1.63 electrons (Sc_3C_2 = +1.33 and Au = +0.3 electrons). Again, only 1.39 is the gained (−) by the Ih-C_{80} cage, the difference of 0.24 electrons gained by C_2 atoms within the moiety. Summing up the two negative figures (1.39 and 0.24), we obtain the total transferred electrons to be 1.63 electrons. Looking at the numbers in brackets mainly Au donation and C_2, N gaining, one could summarise that Au donation is approximately 0.2–0.3 and C2 and N gaining 0.24–1.60 electrons. Furthermore, in all cases, the gain by C_2 and N is larger than Au donation; we attribute that to the fact that C_2 and N atoms are in direct contact with EMF moieties.

Again, in the erbium nitride configuration ($Sc_3N@C_{80}$ + Au), net charge transfers are significantly larger than those of the scandium nitride and scandium carbide configurations. Tables 1 and 2 show that the net charge transfers of Er_3N are more than four times higher than those of Sc_3C_2 and Sc_3N, and the reason for this is that the erbium nitride moiety possesses f-electrons in its outer orbital shells [30]. It is widely known that DFT cannot treat electrons in f-orbitals accurately [31].

The above result explains why the $\Delta(\theta)$ of Er_3N is less symmetric than that of Sc_3C_2 and Sc_3N, as shown in Figure 3 (Note: Figure 3 is reported in our previous work [29]). This also applies to the charge inhomogeneity (σ_q). In ref. [29], the standard deviations of charge distributions on the three EMFs complexes were evaluated, and the values indicated that the σ_q of the erbium nitride complex was 10 times less than that of the other complexes, as illustrated in Table 3.

Table 3. Standard deviations of charge σ_q for $Sc_3C_2@C_{80}$, $Sc_3N@C_{80}$, and $Er_3N@C_{80}$ complexes. Charges are simulated using the Mulliken, Hirshfeld, and Voronoi methods. Adapted with permission from ref [29]. Copyright 2022 Nanoscale Horizons, 2022.

EMF Complex	$\sigma_{Mulliken}$	$\sigma_{Hirshfeld}$	$\sigma_{Voronoi}$
$Sc_3C_2@C_{80}$	0.0154	0.0113	0.0133
$Sc_3N@C_{80}$	0.0163	0.0109	0.0119
$Er_3N@C_{80}$	0.00378	0.00259	0.00268

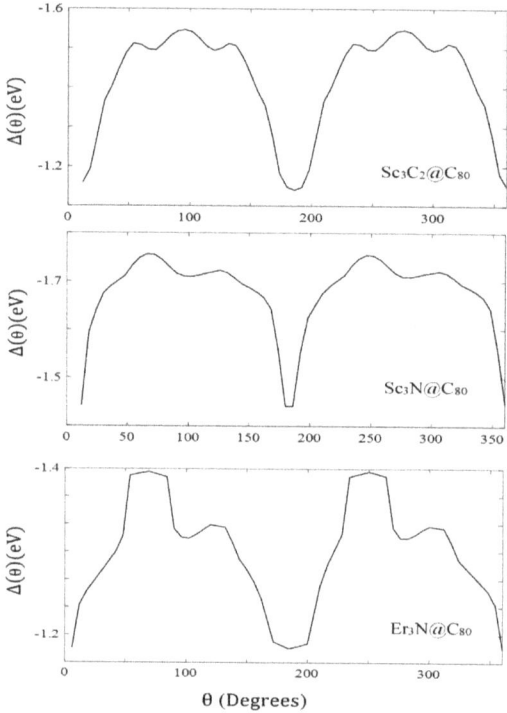

Figure 3. $\Delta(\theta)$ of Er_3N, Sc_3C_2, and Sc_3N within the fullerene cage. Energy barriers $\Delta E(\theta)$ to rotation about θ. Adapted with permission from [29]. Copyright 2022 Nanoscale Horizons, 2022.

CT analyses that were performed in both the solute and on a Au surface are essential to determine the total number of electrons transferred from metallic moieties to the Ih-C_{80} cage. These charge transfers have a great influence on the electric and thermoelectric properties of the EMFs. The electrical conductance is boosted in crown ether molecules [32,33], due to charge transfer from the ion to the molecular wire, causing the molecular resonances to shift closer to the electrode Fermi energy. Similarly, CT enhances the Seebeck coefficient in crown ether molecules [34–37] and endohedral metallofullerenes [38–40].

Many experimental and theoretical studies pointed out that EMF–complex junctions retain a high single molecule power factor. For example, Lee and his co-workers [41] reported that the $Gd@C_{82}$ complex has the biggest power factor for a molecular device (at the time of publication), which is about 16.2 fW K^{-2}. This is equivalent to approximately 4×10 µW K^{-2} m^{-1} for a $Gd@C_{82}$ monolayer. In another study [29] performed in 2022, the researchers noticed a larger PF of 50 fW K^{-2} for $Sc_3N@C_{80}$ and $Sc_3C_2@C_{80}$ complexes, and some measurements hit 70–80 fW K^{-2} for $Sc_3N@C_{80}$ and $Sc_3C_2@C_{80}$ complexes. Statistically, they report larger values for the carbide complex ($Sc_3C_2@C_{80}$). Considering all their measured conductance and Seebeck coefficient values, the PF can be statistically improved when the charge transfer becomes larger. We attribute this desirable feature to the CT phenomena. Table 3 above clearly illustrates that the charge transfer of $Sc_3N@C_{80}$ and $Sc_3C_2@C_{80}$ complexes are approximately 10 times larger than that of the $Er_3N@C_{80}$ complex, and this explains why their conductance G and Seebeck S (i.e., power factor GS^2) are larger than those of $Er_3N@C_{80}$, as shown in Figure 4 below (Note: Figure 4 has been reported in our previous work [29]).

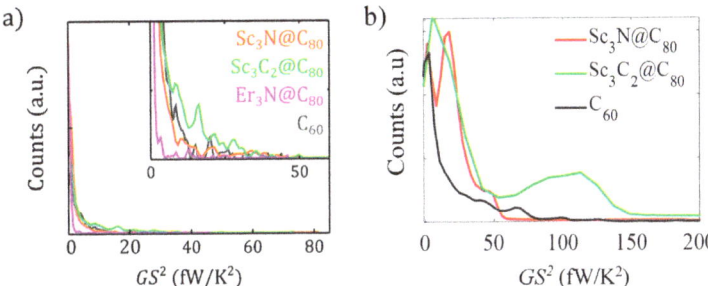

Figure 4. GS^2 analysis. (**a**) Experimental histograms of PF at first contact, built with the data in Figure 3. The inset zooms into the details of the main panel. (**b**). Theoretical 1D histograms of power factor obtained from Supplementary Figures S14 and S15 of the Supplementary Materials. Adapted with permission from [29]. Copyright 2022 Nanoscale Horizons, 2022.

Large power factor GS^2 and bi-thermoelectric behaviour of the studied EMF complexes underline our initial intuition that charge transfer (CT) results in considerable improvement in the thermoelectric transport properties, compared with an empty cage such as C_{60}. This desirable feature has also been noted previously [41] in some EMF complexes involving Gd@C_{82} and Ce@C_{82}, and the empty C_{82} displayed mainly negative Seebeck coefficients, with occasional positive Seebeck coefficients. The positive and negative Seebeck coefficients were ascribed to meta-geometries. The reported findings of the thermopower demonstrate improvements in the EMFs thermoelectric properties compared to the empty C_{82}. Compared to the current investigated EMF complexes, the only difference with the complexes explored in a previous study [41] is the number of metallic atoms within the complex, and in [41], a single atom was positioned out within the cage. This difference could lead only to less migrated charge to the fullerene cage as the metallic moiety is smaller.

4. Conclusions

In conclusion, through a systematic theory study, we have demonstrated that the electrical and thermoelectrical performance of endohedral metallofullerene (EMF) complexes and configurations can be modulated by chemically varying the metallic moiety that encapsulates inside Ih-C_{80} cages. The electric charge transfer of three EMFs involving Sc_3C_2@C_{80}, Sc_3N@C_{80}, and Er_3N@C_{80} complexes and their configurations when they are placed on a gold (111) surface have been investigated in three different charge transfer methods. The Mulliken, Hirshfeld, and Voronoi methods all suggest that the charge migrates from metallic moieties such as Sc_3C_2, Sc_3N, and Er_3N to C_{80} cages; however, the amount of the transferred charge depends on the nature of the moiety inside the EMF complex. Published studies [32,33,38,42–45] evidenced that the CT improve both conductance and thermopower. This work sheds light on new strategies for designing electric and thermoelectric devices based on tuning the CT by using different metallic moieties with potential practical applications.

Supplementary Materials: The following supporting information can be downloaded at: https://www.mdpi.com/article/10.3390/en16114342/s1, Figure S1: Geometries of an asymmetric Sc3C2 (a), symmetric Sc3N and Er3N moieties (b,c). Key: C = grey, N = blue and Er = O = red; Figure S2: Endohedral metallofullerenes and fullerene studied Molecules. Schematic of the three endohedral metallofullerenes (EMFs), namely, a: Sc_3C_2@C_{80}, b: Sc_3N@C_{80}, and c: Er_3N@C_{80} and an empty fullerene cage d: C_{80}; Figure S3: Wave function plots of Sc_3C_2@C_{80} complex. Top panel: fully optimised geometry of Sc_3C_2@C_{80} EMF. Lower panel: HOMO, LUMO, HOMO-1, LUMO+1 of Sc_3C_2@C_{80} complex along with their energies; Figure S4: Wave function plots of Er_3N@C_{80} complex. Top panel: fully optimised geometry of Sc_3C_2@C_{80} EMF. Lower panel: HOMO, LUMO, HOMO-1, LUMO+1 of Er_3N@C_{80} complex along with their energies; Figure S5: Wave function plots of Er3N@C80 complex. Top panel: fully optimised geometry of Sc3C2@C80 EMF. Lower panel: HOMO,

LUMO, HOMO-1, LUMO+1 of Er3N@C80 complex along with their energies; Figure S6: Sc3C2@C80 on a gold surface (Right panel). Energy difference of Sc3C2@C80 /gold complex as a function of molecule-gold distance. The equilibrium distance corresponding to the energy minimum is found to be approximately 2.5\Å (Left panel); Figure S7: Seebeck coefficient S as a function of Fermi energy at 60 different orientations angles \theta of Sc3C2@C80, for a tip-substrate distance of 2.5 Å; Figure S8: Seebeck coefficients S as a function of Fermi energy at 60 different orientation angles\theta of Sc3N@C80 for a tip-substrate distance of 2.5 Å. Table S1: Charge transfer analyses using Mulliken, Hirshfeld and Voronoi methods of $Sc_3C_2@C_{80}$, $Sc_3N@C_{80}$ and $Er_3N@C_{80}$ complexes. The total number of electrons transferred from metallic moieties (with a charge of $+|e|$), to Ih-C_{80} cages (with a charge of $-|e|$), to form complexes. Note: loss-gain differences gain by C_2, N and N (numbers in brackets), $Sc_3C_2@C_{80}$, $Sc_3N@C_{80}$ and $Er_3N@C_{80}$ complexes in gas phase; Table S2: Charge transfer analyses using Mulliken, Hirshfeld and Voronoi methods of $Sc_3C_2@C_{80}$, $Sc_3N@C_{80}$ and $Er_3N@C_{80}$ complexes. The total number of electrons transferred from metallic moieties (with a charge of $+|e|$), to Ih-C_{80} cages (with a charge of $-|e|$), to form complexes. Note: loss-gain differences gain by C_2, N and N (numbers in brackets), of $Sc_3C_2@C_{80}$, $Sc_3N@C_{80}$ and $Er_3N@C_{80}$ complexes on an Au (111), surface.

Author Contributions: A.K.I. originally conceived the concept; calculations were carried out by M.A. (Majed Alshammari), M.A. (Moteb Altoaibi) and T.A. All authors provided essential contributions to interpreting the data reported in this manuscript. A.K.I. coordinated the writing of the manuscript with input from M.A. (Majed Alshammari), M.A. (Moteb Altoaibi) and T.A. All authors have read and agreed to the published version of the manuscript.

Funding: This work was supported by the Leverhulme Trust for Early Career Fellowship ECF-2020-638. This work was additionally funded by the European Commission FET Open projects 767187-QuIET and 766853-EFINED. M.A. (Majed Alshammari) and T.A. are grateful for the financial assistance from Jouf University (Saudi Arabia), M.A. (Majed Alshammari) and T.A. are thankful for computer time, this research used the resources of the Supercomputing Laboratory at King Abdullah University of Science & Technology (KAUST) in Thuwal, Saudi Arabia. M.A. (Moteb Altoaibi) is grateful for the sported the Deanship of Scientific Research at Prince Sattam bin Abdulaziz University, Alkharj, SaudiArabia and the Saudi Ministry of Education. A.K.I. is grateful for financial assistance from Tikrit University (Iraq), and the Iraqi Ministry of Higher Education (SL-20).

Conflicts of Interest: The authors declare no conflict of interest.

References

1. Soos, Z.G. Theory of π-molecular charge-transfer crystals. *Annu. Rev. Phys. Chem.* **1974**, *25*, 121–153. [CrossRef]
2. Bauer, C.; Teuscher, J.; Brauer, J.C.; Punzi, A.; Marchioro, A.; Ghadiri, E.; De Jonghe, J.; Wielopolski, M.; Banerji, N.; Moser, J.-E. Dynamics and mechanisms of interfacial photoinduced electron transfer processes of third generation photovoltaics and photocatalysis. *CHIMIA Int. J. Chem.* **2011**, *65*, 704–709. [CrossRef]
3. Günes, S.; Neugebauer, H.; Sariciftci, N.S. Conjugated polymer-based organic solar cells. *Chem. Rev.* **2007**, *107*, 1324–1338. [CrossRef]
4. Megiatto, J.D., Jr.; Méndez-Hernández, D.D.; Tejeda-Ferrari, M.E.; Teillout, A.-L.; Llansola-Portolés, M.J.; Kodis, G.; Poluektov, O.G.; Rajh, T.; Mujica, V.; Groy, T.L. A bioinspired redox relay that mimics radical interactions of the Tyr–His pairs of photosystem. *Nat. Chem.* **2014**, *6*, 423–428. [CrossRef] [PubMed]
5. Aviram, A.; Ratner, M.A. Molecular rectifiers. *Bull. Am. Phys. Soc.* **1974**, *19*, 341. [CrossRef]
6. Herrer, L.; Ismael, A.; Martin, S.; Milan, D.C.; Serrano, J.L.; Nichols, R.J.; Lambert, C.; Cea, P. Single molecule vs. large area design of molecular electronic devices incorporating an efficient 2-aminepyridine double anchoring group. *Nanoscale* **2019**, *11*, 15871–15880. [CrossRef]
7. Al-Khaykanee, M.K.; Ismael, A.K.; Grace, I.; Lambert, C.J. Oscillating Seebeck coefficients in π-stacked molecular junctions. *Rsc Adv.* **2018**, *8*, 24711–24715. [CrossRef]
8. Bockrath, M.; Cobden, D.H.; McEuen, P.L.; Chopra, N.G.; Zettl, A.; Thess, A.; Smalley, R.E. Single-electron transport in ropes of carbon nanotubes. *Science* **1997**, *275*, 1922–1925. [CrossRef]
9. Ismael, A.K.; Lambert, C.J. Single-molecule conductance oscillations in alkane rings. *J. Mater. Chem. C* **2019**, *7*, 6578–6581. [CrossRef]
10. Romaner, L.; Heimel, G.; Brédas, J.-L.; Gerlach, A.; Schreiber, F.; Johnson, R.L.; Zegenhagen, J.; Duhm, S.; Koch, N.; Zojer, E. Impact of bidirectional charge transfer and molecular distortions on the electronic structure of a metal-organic interface. *Phys. Rev. Lett.* **2007**, *99*, 256801. [CrossRef]

11. Bennett, T.L.; Alshammari, M.; Au-Yong, S.; Almutlg, A.; Wang, X.; Wilkinson, L.A.; Albrecht, T.; Jarvis, S.P.; Cohen, L.F.; Ismael, A. Multi-component self-assembled molecular-electronic films: Towards new high-performance thermoelectric systems. *Chem. Sci.* **2022**, *13*, 5176–5185. [CrossRef] [PubMed]
12. Lu, D.; Chen, G.; Perry, J.W.; Goddard, W.A., III. Valence-bond charge-transfer model for nonlinear optical properties of charge-transfer organic molecules. *J. Am. Chem. Soc.* **1994**, *116*, 10679–10685. [CrossRef]
13. Gorczak, N.; Renaud, N.; Tarkuç, S.; Houtepen, A.J.; Eelkema, R.; Siebbeles, L.D.; Grozema, F.C. Charge transfer versus molecular conductance: Molecular orbital symmetry turns quantum interference rules upside down. *Chem. Sci.* **2015**, *6*, 4196–4206. [CrossRef] [PubMed]
14. Closs, G.L.; Miller, J.R. Intramolecular long-distance electron transfer in organic molecules. *Science* **1988**, *240*, 440–447. [CrossRef]
15. Sukegawa, J.; Schubert, C.; Zhu, X.; Tsuji, H.; Guldi, D.M.; Nakamura, E. Electron transfer through rigid organic molecular wires enhanced by electronic and electron–vibration coupling. *Nat. Chem.* **2014**, *6*, 899–905. [CrossRef]
16. Deibel, C.; Strobel, T.; Dyakonov, V. Role of the charge transfer state in organic donor–acceptor solar cells. *Adv. Mater.* **2010**, *22*, 4097–4111. [CrossRef]
17. Otero, R.; de Parga, A.V.; Gallego, J.M. Electronic, structural and chemical effects of charge-transfer at organic/inorganic interfaces. *Surf. Sci. Rep.* **2017**, *72*, 105–145. [CrossRef]
18. Kollmannsberger, M.; Rurack, K.; Resch-Genger, U.; Rettig, W.; Daub, J. Design of an efficient charge-transfer processing molecular system containing a weak electron donor: Spectroscopic and redox properties and cation-induced fluorescence enhancement. *Chem. Phys. Lett.* **2000**, *329*, 363–369. [CrossRef]
19. Wörner, H.J.; Arrell, C.A.; Banerji, N.; Cannizzo, A.; Chergui, M.; Das, A.K.; Hamm, P.; Keller, U.; Kraus, P.M.; Liberatore, E. Charge migration and charge transfer in molecular systems. *Struct. Dyn.* **2017**, *4*, 061508. [CrossRef]
20. Mulliken, R.S. Electronic population analysis on LCAO–MO molecular wave functions. *J. Chem. Phys.* **1955**, *23*, 1833–1840. [CrossRef]
21. Hirshfeld, F.L. Bonded-atom fragments for describing molecular charge densities. *Theor. Chim. Acta* **1977**, *44*, 129–138. [CrossRef]
22. Guerra, C.F.; Handgraaf, J.W.; Baerends, E.J.; Bickelhaupt, F.M. Voronoi deformation density (VDD) charges: Assessment of the Mulliken, Bader, Hirshfeld, Weinhold, and VDD methods for charge analysis. *J. Comput. Chem.* **2004**, *25*, 189–210. [CrossRef]
23. Soler, J.M.; Artacho, E.; Gale, J.D.; García, A.; Junquera, J.; Ordejón, P.; Sánchez-Portal, D.J.J.o.P.C.M. The SIESTA method for ab initio order-N materials simulation. *J. Phys. Condens. Matter* **2002**, *14*, 2745. [CrossRef]
24. Davidson, R.J.; Milan, D.C.; Al-Owaedi, O.A.; Ismael, A.K.; Nichols, R.J.; Higgins, S.J.; Lambert, C.J.; Yufit, D.S.; Beeby, A. Conductance of 'bare-bones' tripodal molecular wires. *RSC Adv.* **2018**, *8*, 23585–23590. [CrossRef] [PubMed]
25. Markin, A.; Ismael, A.K.; Davidson, R.J.; Milan, D.C.; Nichols, R.J.; Higgins, S.J.; Lambert, C.J.; Hsu, Y.-T.; Yufit, D.S.; Beeby, A. Conductance Behavior of Tetraphenyl-Aza-BODIPYs. *J. Phys. Chem. C* **2020**, *124*, 6479–6485. [CrossRef]
26. Kobko, N.; Dannenberg, J. Dannenberg. Effect of basis set superposition error (BSSE) upon ab initio calculations of organic transition states. *J. Phys. Chem. A* **2001**, *105*, 1944–1950. [CrossRef]
27. Sherrill, C.D. *Counterpoise Correction and Basis Set Superposition Error*; School of Chemistry and Biochemistry, Georgia Institute of Technology: Atlanta, Georgia, 2010.
28. Sinnokrot, M.O.; Valeev, E.F.; Sherrill, C.D. Estimates of the ab initio limit for $\pi-\pi$ interactions: The benzene dimer. *J. Am. Chem. Soc.* **2002**, *124*, 10887–10893. [CrossRef]
29. Ismael, A.K.; Rincón-García, L.; Evangeli, C.; Dallas, P.; Alotaibi, T.; Al-Jobory, A.A.; Rubio-Bollinger, G.; Porfyrakis, K.; Agraït, N.; Lambert, C.J. Exploring seebeck-coefficient fluctuations in endohedral-fullerene, single-molecule junctions. *Nanoscale Horiz.* **2022**, *7*, 616–625. [CrossRef]
30. Akkermans, E.; Montambaux, G. *Mesoscopic Physics of Electrons and Photons*; Cambridge University Press: Cambridge, UK, 2007. [CrossRef]
31. Cohen, A.J.; Mori-Sánchez, P.; Yang, W. Challenges for density functional theory. *Chem. Rev.* **2012**, *112*, 289–320. [CrossRef]
32. Ismael, A.K.; Al-Jobory, A.; Grace, I.; Lambert, C.J. Discriminating single-molecule sensing by crown-ether-based molecular junctions. *J. Chem. Phys.* **2017**, *146*, 064704. [CrossRef]
33. Ismael, A.K.; Grace, I.; Lambert, C.J. Increasing the thermopower of crown-ether-bridged anthraquinones. *Nanoscale* **2015**, *7*, 17338–17342. [CrossRef] [PubMed]
34. Ismael, A.K.; Grace, I.; Lambert, C.J. Connectivity dependence of Fano resonances in single molecules. *Phys. Chem. Chem. Phys.* **2017**, *19*, 6416–6421. [CrossRef] [PubMed]
35. Wang, X.; Ismael, A.; Ning, S.; Althobaiti, H.; Al-Jobory, A.; Girovsky, J.; Astier, H.P.; O'Driscoll, L.J.; Bryce, M.R.; Lambert, C.J. Electrostatic Fermi level tuning in large-scale self-assembled monolayers of oligo (phenylene–ethynylene) derivatives. *Nanoscale Horiz.* **2022**, *7*, 1201–1209. [CrossRef] [PubMed]
36. Wilkinson, L.A.; Bennett, T.L.; Grace, I.M.; Hamill, J.; Wang, X.; Au-Yong, S.; Ismael, A.; Jarvis, S.P.; Hou, S.; Albrecht, T. Assembly, structure and thermoelectric properties of 1,1'-dialkynylferrocene 'hinges'. *Chem. Sci.* **2022**, *13*, 8380–8387. [CrossRef] [PubMed]
37. Ye, J.; Al-Jobory, A.; Zhang, Q.-C.; Cao, W.; Alshehab, A.; Qu, K.; Alotaibi, T.; Chen, H.; Liu, J.; Ismael, A.K. Highly insulating alkane rings with destructive σ-interference. *Sci. China Chem.* **2022**, *65*, 1822–1828. [CrossRef]
38. Rincón-García, L.; Ismael, A.K.; Evangeli, C.; Grace, I.; Rubio-Bollinger, G.; Porfyrakis, K.; Agraït, N.; Lambert, C.J. Molecular design and control of fullerene-based bi-thermoelectric materials. *Nat. Mater.* **2016**, *15*, 289–293. [CrossRef]

39. Lu, J.; Nagase, S.; Zhang, X.; Wang, D.; Ni, M.; Maeda, Y.; Wakahara, T.; Nakahodo, T.; Tsuchiya, T.; Akasaka, T. Selective interaction of large or charge-transfer aromatic molecules with metallic single-wall carbon nanotubes: Critical role of the molecular size and orientation. *J. Am. Chem. Soc.* **2006**, *128*, 5114–5118. [CrossRef]
40. Ismael, A.; Al-Jobory, A.; Wang, X.; Alshehab, A.; Almutlg, A.; Alshammari, M.; Grace, I.; Benett, T.L.; Wilkinson, L.A.; Robinson, B.J. Molecular-scale thermoelectricity: As simple as 'ABC'. *Nanoscale Adv.* **2020**, *2*, 5329–5334. [CrossRef]
41. Lee, S.K.; Buerkle, M.; Yamada, R.; Asai, Y.; Tada, H. Thermoelectricity at the molecular scale: A large Seebeck effect in endohedral metallofullerenes. *Nanoscale* **2015**, *7*, 20497–20502. [CrossRef]
42. Balachandran, J.; Reddy, P.; Dunietz, B.D.; Gavini, V. End-group-induced charge transfer in molecular junctions: Effect on electronic-structure and thermopower. *J. Phys. Chem. Lett.* **2012**, *3*, 1962–1967. [CrossRef]
43. Adams, D.M.; Brus, L.; Chidsey, C.E.; Creager, S.; Creutz, C.; Kagan, C.R.; Kamat, P.V.; Lieberman, M.; Lindsay, S.; Marcus, R.A. Charge transfer on the nanoscale: Current status. *J. Phys. Chem. B* **2003**, *107*, 6668–6697. [CrossRef]
44. Liu, S.-X.; Ismael, A.K.; Al-Jobory, A.; Lambert, C.J. Signatures of Room-Temperature Quantum Interference in Molecular Junctions. *Acc. Chem. Res.* **2023**, 4193–4201. [CrossRef] [PubMed]
45. Alshehab, A.; Ismael, A.K. Impact of the terminal end-group on the electrical conductance in alkane linear chains. *RSC Adv.* **2023**, *13*, 5869–5873. [CrossRef] [PubMed]

Disclaimer/Publisher's Note: The statements, opinions and data contained in all publications are solely those of the individual author(s) and contributor(s) and not of MDPI and/or the editor(s). MDPI and/or the editor(s) disclaim responsibility for any injury to people or property resulting from any ideas, methods, instructions or products referred to in the content.

Review

Heat Transfer Mechanisms and Contributions of Wearable Thermoelectrics to Personal Thermal Management

Diana Enescu [1,2]

1. Electronics, Telecommunications and Energy Department, University Valahia of Targoviste, 130004 Targoviste, Romania; diana.enescu@valahia.ro or d.enescu@inrim.it
2. Istituto Nazionale di Ricerca Metrologica (INRiM), 10135 Torino, Italy

Abstract: Thermoelectricity can assist in creating comfortable thermal environments through wearable solutions and local applications that keep the temperature comfortable around individuals. In the analysis of an indoor environment, thermal comfort depends on the global characteristics of the indoor volume and on the local thermal environment where the individuals develop their activity. This paper addresses the heat transfer mechanisms that refer to individuals, which operate in their working ambient when wearable thermoelectric solutions are used for enhancing heating or cooling within the local environment. After recalling the characteristics of the thermoelectric generators and illustrating the heat transfer mechanisms between the human body and the environment, the interactions between wearable thermoelectric generators and the human skin are discussed, considering the analytical representations of the thermal phenomena. The wearable solutions with thermoelectric generators for personal thermal management are then categorized by considering active and passive thermal management methods, natural and assisted heat exchange, autonomous and nonautonomous devices, and direct or indirect contact with the human body.

Keywords: wearable thermoelectric generator; thermal comfort; local thermal environment; convective heat transfer coefficient; thermal sensation; personal thermal management

Citation: Enescu, D. Heat Transfer Mechanisms and Contributions of Wearable Thermoelectrics to Personal Thermal Management. *Energies* **2024**, *17*, 285. https://doi.org/10.3390/en17020285

Academic Editor: Wei-Hsin Chen

Received: 27 November 2023
Revised: 26 December 2023
Accepted: 29 December 2023
Published: 5 January 2024

Copyright: © 2024 by the author. Licensee MDPI, Basel, Switzerland. This article is an open access article distributed under the terms and conditions of the Creative Commons Attribution (CC BY) license (https://creativecommons.org/licenses/by/4.0/).

1. Introduction

A thermoelectric generator (TEG) is a solid-state device that operates due to the Seebeck phenomenon to harvest electrical energy by converting the temperature difference ΔT determined by human body heat (considered as a heat source) and surrounding conditions into voltage ΔV [1]. The parameter called the Seebeck coefficient or thermopower is $S = -\frac{\Delta V}{\Delta T}$. The TEG device generates electricity in direct current (DC) as long as there is a temperature difference between its sides. In this way, the TEG operates due to the movement of the charge carriers (electrons and holes) within one or more thermoelement pairs (or thermocouples) connected together. The two thermoelements that form the pair are made of semiconductor materials (N-type and P-type) connected to each other at one end through a metallic strip (or copper interconnect or metal electrodes), forming a junction (as shown in Figure 1). When the junction is subject to a temperature gradient, the heat carriers (electrons and holes) begin to move through the two thermoelements. In the N-type thermoelement of a TEG, there is an excess of negative charge carriers, i.e., electrons. These electrons are set in motion by the temperature gradient and migrate from the hotter side to the colder side, creating an electrical potential. In the P-type thermoelement, there is an excess of positive charge carriers, i.e., holes, which also move in response to the temperature gradient, generating an electrical potential. Both electrons and holes move in the same direction under the same temperature gradient, and this results in the creation of a difference of potential at the TEG terminals.

The thermoelements are arranged in a regular matrix inside the TEG and are connected electrically in series and thermally in parallel. In addition, the thermocouple is placed between two ceramic plates or ceramic substrates. The ceramic plates are excellent electrical insulators and maintain the thermoelements insulated from the electrical point of view. At the same time, the ceramic plates are good conductors from a thermal point of view. In addition, the ceramic plates serve as mechanical support upon which the thermoelements are mounted. The common ceramic plate is made of aluminum oxide (Al_2O_3). A temperature gradient between the ceramic plates leads to a temperature difference along the thermoelements. Keeping an appropriate temperature gradient along the thermoelements, a heat sink is usually attached to the cold side of the TEGs to speed up the heat dissipation. The higher output voltage and electric power are obtained by rising ΔT between the heat source and the heat sink.

An electrical load having resistance R_L can be connected to the output terminals of TEG, creating an electric circuit in which there is the circulation of a current.

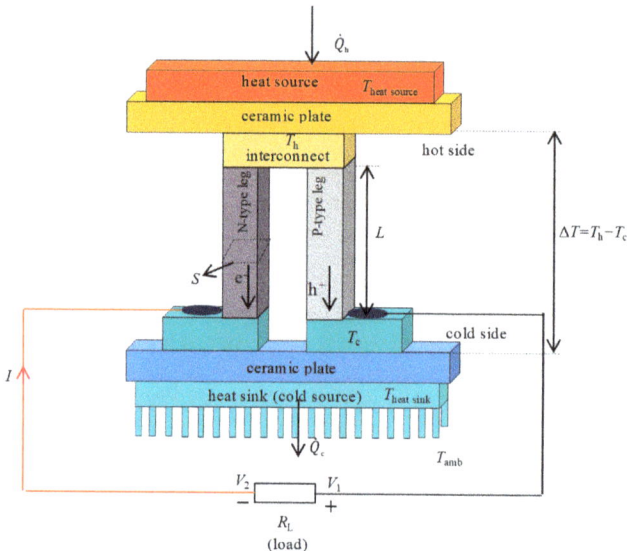

Figure 1. Sketch of the traditional TEG device ($\dot{Q}_h$—the incoming heat flow at the hot side of TEG; $\dot{Q}_c$—the outcoming heat flow at the cold side of TEG) [2].

The efficiency of the thermoelectric material is gauged by the dimensionless figure of merit $Z\overline{T}$, which depends on the three transport properties of the material (electrical conductivity σ, thermal conductivity k, and Seebeck coefficient S), as well as the mean absolute temperature $\overline{T}$, and is given by the following equation:

$$Z\overline{T} = \frac{S^2 \cdot \sigma \cdot \overline{T}}{k} \quad (1)$$

The mean absolute temperature $\overline{T}$ is the arithmetic mean between the temperature T_h at the hot side and the temperature T_c at the cold side.

The product $S^2\sigma$ is the power factor, representing the main parameter to evaluate the performance of the thermoelectric materials. A thermoelectric material requires high-power factor (high S and high σ), and low k. The thermal conductivity of a semiconductor depends on the charge carriers' thermal conductivity and the lattice's thermal conductivity:

$$k = k_{\text{charge carrier}} + k_{\text{lattice}} \quad (2)$$

The charge carrier thermal conductivity depends on the Lorenz constant L, the electrical conductivity σ and the absolute temperature according to the Wiedemann–Franz law:

$$k_{\text{charge carrier}} = L \cdot \sigma \cdot T \qquad (3)$$

The thermal conductivity of the lattice is given by:

$$k_{\text{lattice}} = a \cdot c_p \cdot \rho \qquad (4)$$

where a is the thermal diffusivity, c_p is the specific heat at constant pressure and ρ is the density.

There is no theoretical limit for the dimensionless figure of merit, but the most efficient semiconductor materials used today have values around unity at room temperature.

The conversion efficiency of the TEGs depends on the dimensionless figure of merit and the temperature difference along the thermoelements, $\Delta T = T_h - T_c$.

The maximum theoretical conversion efficiency of a TEG is given by:

$$\eta_{\max} = \left(\frac{T_h - T_c}{T_h}\right) \cdot \frac{\sqrt{1 + Z\overline{T}} - 1}{\sqrt{1 + Z\overline{T}} + \frac{T_c}{T_h}} = \eta_C \cdot \frac{\sqrt{1 + Z\overline{T}} - 1}{\sqrt{1 + Z\overline{T}} + \frac{T_c}{T_h}} \qquad (5)$$

where η_C is the Carnot efficiency, which represents a superior limit on using waste heat for thermoelectric power generation [3]. The maximum theoretical conversion efficiency of the TEG is directly influenced by ΔT. Raising ΔT along the thermoelements, the maximum conversion efficiency of the TEG rises. To regulate η_C, one of the most encouraging approaches is the analysis of the structural design of the thermoelements with respect to the optimal ratio of the cross-sectional area of the P-type (S_P) and N-type (S_N) thermoelements [4]. The optimal ratio is given by:

$$\frac{S_N}{S_P} = \sqrt{\frac{\sigma_P \cdot k_P}{\sigma_N \cdot k_N}} \qquad (6)$$

Structural design classifies TEGs into two categories: flexible TEGs and rigid TEGs. The rigid TEGs have a sandwich structure, while flexible TEGs feature a thin-film structure or fabric-based construction. Furthermore, rigid TEGs are used in powering medical equipment, while flexible TEGs are used for powering wearable equipment [4,5]. In the case of rigid TEGs, the thermoelements are linked using copper strips and are placed between ceramic plates such as alumina (with a thermal conductivity of approximately 30 W/(m·K) or aluminum nitride (with thermal conductivity of approximately 285 W/(m·K)) [6]. For the flexible TEGs, the rigid substrates are eliminated, and the gap between thermoelements is filled with filler material (e.g., flexible polymers with very low thermal conductivity, while the exposed copper strips are into contact with the body's skin [4]). In flexible TEGs, a liquid metal interconnect made of a gallium and indium eutectic alloy (EGaIn) has been used, providing both flexibility and self-healing properties that maintain the TEG module's integrity even after enduring significant strains [7]. The two most used flexible substrates are polydimethylsiloxane (PDMS) and Kapton HN. However, compared to rigid substrates, flexible substrates can hinder the transfer of heat from the human body to the TEG due to their higher thermal resistance [8]. The flexible TEG has both a cross-plane structural design and an in-plane structural design, while rigid TEG has a cross-plane structural design [1,9]:

- In the cross-plane structural design, the thermoelements are placed perpendicularly to the substrate. For a curved surface, the cross-plane structural design has a high flexibility degree, which can bend up to 2 mm in radius for a curved surface. For this reason, the cross-plane structural design is more adequate for human body energy applications than the in-plan structural design.

○ In the in-plane structural design, the thermoelements are parallel to the substrate. Due to the low ΔT between the hot and cold sides in the cross-plane structural design, this design cannot generate more power.

The TEG device has several benefits considering it is portable, noiseless, and requires no maintenance [10]. These advantages are also useful for a wearable TEG (w-TEG), which converts the heat generated by the human body into electricity. The w-TEG is used as a battery booster or as a single power source in many wearable electronic devices (e.g., multisensory electronic skin [11], glucose sensors [12], and electrocardiogram sensors [13,14]).

The design and fabrication of TEG devices are limited by the type of material. The utilization of inorganic materials (e.g., bismuth telluride (Bi_2Te_3) [15,16] and antimony telluride (Sb_2Te_3) [17,18], lead telluride (PbTe), and germanium telluride (GeTe). Even if these materials have high performance with superior stability at ambient temperature, they are mechanically rigid and require advanced techniques to manufacture the wanted thermoelement structure [19]. For this reason, it is difficult enough to integrate inorganic semiconductors into flexible w-TEGs for harvesting energy from irregular surfaces (e.g., the human wrist) [13]. Inorganic semiconductors are toxic, scarce, and expensive, with relatively poor processability, limiting their large applications. In addition, inorganic semiconductors have high thermal conductivity from about 1.2 to 1.6 W/(m·K) [20]. Conversely, organic materials, recently exploited in power generation applications, are considered good candidates due to their light weight, flexibility, processability, low-cost manufacturing, and abundant raw material. In addition, organic materials have a thermal conductivity of around 0.5 W/(m·K), being very close to the lower limit of inorganic materials [21,22].

This paper contains a review of the application of w-TEG devices for personal thermal management. Unlike other review papers, which have focused on thermoelectric technologies and applications with details on specific materials and devices, this paper takes a different approach. Starting with the heat transfer principles that underlie the interaction between thermoelectric devices and the human body, this paper discusses the implications of using w-TEG for local thermal comfort. Specific applications of wearable thermoelectric devices for monitoring the biophysical parameters [23] or to optimize the design of thermoelectric devices [24] are outside the scope of this paper.

The next sections of this paper are organized as follows: Section 2 discusses the heat transfer mechanisms between the human body and the environment, which are important for understanding how to use the human body as a heat source for w-TEGs. Section 3 addresses the interactions between w-TEG and the human skin, recalling the thermal analytical representations. Section 4 summarizes the use of w-TEGs for personal thermal management. The last section contains the conclusions.

2. Basic Principles of Human Body Dry Heat Transfer to the Environment: Harnessing the Human Body as a Heat Source for Wearable TEGs

2.1. Heat Transfer Mechanisms

Heat transfer from the human body to the surroundings is vital for the human body's thermoregulation. The human thermoregulatory system utilizes a range of coordinated physiological mechanisms. These mechanisms are:

- Insulation, which slows down the transfer of heat from the body, helping maintain a comfortable internal temperature.
- Sweating, which cools the body through the evaporation of water.
- Shivering, which generates heat via muscle contraction.
- Vasodilation and vasoconstriction, which regulate blood flow and distribute heat throughout the body.

When the body's core temperature rises, mechanisms such as sweating and vasodilation are activated to cool the body. Conversely, when the core temperature falls, the system initiates mechanisms like shivering and vasoconstriction that warm the body [25]. The physiological mechanisms serve as the functional tools the body deploys to execute its thermoregulatory responses.

The human body continuously produces heat through metabolic processes—the chemical reactions within our cells. This metabolic heat is essential to sustain the body's core temperature, which, for optimal function, should remain around 37 °C (homeothermy). Metabolic heat, also known as metabolic rate, is measured in metabolic equivalents (Met). For example, a sleeping person's metabolism is roughly 46 W/m² (0.8 Met), a person sitting in an office has around 70 W/m² (1.2 Met), and during intense physical activity, the metabolic rate can spike up to 550 W/m² (9.5 Met) [26]. Metabolic heat production in the human body directly influences the subsequent heat transfer processes, as it determines the amount of heat that needs to be dissipated to maintain the core temperature. The body utilizes several processes for heat transfer, including conduction, convection, radiation, and evaporation. Understanding these mechanisms is crucial, especially in fields like garment design and safety gear. These insights also influence the design of environmental control systems, such as heating, ventilation and air conditioning systems, and lighting, ensuring optimal comfort for the occupants [21].

2.2. Conductive Heat Transfer

Conductive heat transfer takes place when a part of the body comes into contact with a solid surface in the environment. Conductive heat exchange depends on the surface area of body parts in direct contact with external surfaces, and it is typically minimal.

In steady-state conditions, the conductive thermal flux density is expressed by the following equation:

$$\dot{q}_{cond} = k \cdot (t_{skin} - t_{surface}) \tag{7}$$

where k is the thermal conductivity in W/(m·K), which is influenced by the thermal properties of both the skin and the solid surface and by the contact solid surface area, t_{skin} is the temperature of the skin body in °C, and $t_{surface}$ is the temperature of the solid surface in °C. In transient conditions, the heat that is exchanged between the skin and the solid surface depends on the thermal inertia of the material. The Merriam-Webster online dictionary defines "thermal inertia" as "the degree of slowness with which the temperature of a body approaches that of its surroundings and which is dependent upon its absorptivity, its specific heat, its thermal conductivity, its dimensions, and other factors" [27]. In other words, thermal inertia expresses how quickly a material can absorb or release heat. With higher thermal inertia, more heat is transferred to the skin or removed from the skin. If there is only a small surface area of the body in contact with another material (e.g., a person who is standing), and that material has low thermal inertia, then a small amount of conductive heat will be transferred. On the other hand, if a higher surface area of the body is in contact with a material (e.g., a person lying down on a surface) and that material has high thermal inertia, then a significant amount of conductive heat will be transferred [21,28].

2.3. Convective Heat Transfer

Convective heat transfer refers to the process of carrying heat away from the skin to the surrounding air. The convective thermal flux density from the skin to the surrounding air is expressed by the following equation:

$$\dot{q}_{conv} = h_c \cdot (\bar{t}_{skin} - t_{surr\ air}) \tag{8}$$

where $t_{surrair}$ is the temperature of the surrounding air, and h_c is the convective heat transfer coefficient, which, according to (8), is the ratio of the convective thermal flux density to the temperature difference from the skin to the surrounding air [29]. The greater the temperature difference between the skin and the surrounding air, the faster the rate of convective heat loss. This is because a larger temperature difference results in a stronger convective current, accelerating the heat transfer process.

For the dressed human body, the convective thermal flux density from the clothes to the surrounding air is expressed by the following equation:

$$\dot{q}_{conv} = f_{clo} \cdot h_c \cdot (\bar{t}_{clo} - t_{surr\ air}) \quad (9)$$

where f_{clo} represents the clothing area factor, which represents the ratio between the clothed body surface area compared to the unclothed body surface area. The clothing area factor is estimated as $f_{clo} = 1 + 0.3 \cdot I_{clo}$, where I_{clo} is the intrinsic clothing ensemble insulation, in clo ($f_{clo} = 1$ for the unclothed body surface area), and $\bar{t}_{clo}$ is the mean surface temperature of the clothed body.

Convective heat transfer from the skin or clothing occurs when moving air perturbs the insulating boundary layer of air around the body's surface. The more rapid the airflow around the body, the narrow the boundary layer on the body's surface, leading to lower thermal insulation for the individual [30].

Convection from the human skin or clothing can be divided into the following types:

- Natural convection occurring for the air velocity $w < 0.2$ m/s;
- Forced convection occurring for the air velocity $w > 1.5$ m/s;
- Mixed-mode convection, which takes place at air velocity $0.2 < w < 1.5$ m/s.

Now, having outlined the three main types of convection, it is important to examine the phenomenon of natural convection and the dimensionless parameter known as the Grashof number. The average skin temperature of a naked individual is higher than the surrounding air temperature. Consequently, the layer of air directly touching the skin surface warms up and, as a result, becomes lighter. In calm conditions, this warm air ascends due to buoyancy forces, and colder air from the surrounding environment flows in to take its place [31]. The upward airflow can exhibit either laminar or turbulent characteristics. In the context of natural convection, a crucial dimensionless parameter that describes this flow is the Grashof number (Gr) [29,31]. This parameter quantifies the relationship between buoyant forces and viscous forces and is defined as follows:

$$Gr = \frac{g \cdot \beta \cdot (t_{skin} - t_{surr\ air}) \cdot \updownarrow^3}{\nu^2} \quad (10)$$

where $g = 9.81$ m/s^2 is the gravitational acceleration; $\beta = \frac{1}{T}$ is the thermal expansion coefficient of the air; $\updownarrow$ is a characteristic length, identified with body height; and ν is the kinematic viscosity of the air.

The flow is laminar if $Gr < 10^9$ [30] and is turbulent for $Gr > 10^{10}$ [32]. In the case of a stationary, unclothed individual, with a temperature difference $\Delta t = \bar{t}_{skin} - t_{surr\ air} \approx 8 \div 10\ °C$, the flow is laminar up to a height of 1 m and turbulent at 1.5 m [32]. Therefore, different flow regimes occur under natural convection and at various height levels, permitting different heat loss rates across body segments.

2.3.1. Natural Convection

Under natural convection ($w < 0.2$ m/s), the convective heat transfer coefficient h_c for the whole human body varies from 3.1 to 5.1 W/(m^2·K), according to different studies [30,33].

Xu et al. [34] analyzed the possible reasons for this variation, including differences in body geometry, body posture, and airflow patterns in the investigated room.

Under natural convection, for a clothed body, the general equation used by many researchers is structured as:

$$h_c = a_n \cdot (\bar{t}_{clo} - t_{surr\ air})^{b_n} \quad (11)$$

where a_n and b_n are the coefficients determined by the specific analyses carried out by the researchers, while for the naked human body the equation is structured as:

$$h_c = a_n \cdot (\bar{t}_{skin} - t_{surr\ air})^{b_n} \quad (12)$$

Table 1 shows the values of the coefficients used in the above equations for clothed or naked human body in various postures.

Table 1. Coefficients for the representations of the convective heat transfer coefficients under natural convection.

Posture	a_n	b_n	w (m/s)	Human Body	Reference
Sitting	2.38	0.25	<0.15	clothed	[35]
Sitting	0.78	0.59	<0.1	naked	[30]
Sitting	1.94	0.23	<0.15	naked	[36]
Sitting (exposed to atmosphere)	1.175	0.351	<0.2	naked	[37]
Sitting (contact with seat)	1.222	0.299	<0.2	naked	[37]
Sitting (cross-legged, floor contact)	1.271	0.355	<0.2	naked	[37]
Sitting (legs out, floor contact)	1.002	0.409	<0.2	naked	[37]
Standing	2.35	1.25	$0.02 < w < 0.1$	naked	[38]
Standing	1.21	0.43	<0.1	naked	[39]
Standing	2.02	0.24	<0.15	naked	[36]
Standing	2.38	0.25	<0.15	clothed	[35]
Standing (exposed to atmosphere)	1.007	0.406	<0.2	naked	[37]
Sitting (floor contact)	1.183	0.347	<0.2	naked	[37]
Supine (floor contact)	0.881	0.368	<0.2	naked	[37]
Lying	2.48	0.18	<0.15	naked	[40]

In other references, only the convective heat transfer coefficients h_c of the human body are indicated. For example, de Dear et al. [30] indicated values in the range from 4 to 6 $\frac{W}{(m^2 \cdot K)}$.

2.3.2. Forced Convection

Under forced convection, numerous efforts have been undertaken throughout the years to empirically determine the convective heat transfer coefficients suitable for the entire human body. The Archimedes number (Ar) plays a key role in determining the relative significance of free and forced convection. Considering the Reynolds number (Re), defined as:

$$Re = \frac{w \cdot \updownarrow}{a} \qquad (13)$$

where $\updownarrow$ is a characteristic length associated with the body and a is the thermal diffusivity, Ar is the ratio between the Grashof number to the Reynolds number squared:

$$Ar = \frac{Gr}{Re^2} \qquad (14)$$

Sparrow et al. [41] provided a criterion for categorizing various types of convective flow based on the Archimedes number. In this case, for forced convection $0 < Ar < 0.3$, for mixed convection $0.3 < Ar < 16$, while for natural convection $Ar > 16$.

In forced convection, where air is blown over the surface, the convective heat transfer coefficient h_c increases with air velocity being a function of it.

For vertical airflow, in the case of the mixed convection, when the air velocity is $w > 0.2$ m/s the following expression holds:

$$h_c = c_f + a_f \cdot w^{b_f} \qquad (15)$$

where a_f, b_f, and c_f are the constants determined by the researchers [42]. For example, Colin and Houdas [43] obtained the values a_f = 8.7, b_f = 0.67, and c_f = 2.7 from the regression for sitting, for the air velocity, in m/s, in the range $0.15 < w < 1.5$.

The convective heat transfer coefficient for downward airflow conditions was greater than that for upward airflow conditions when the air velocity was below 0.3 m/s. However, as the air velocity surpassed 0.3 m/s, the convective heat transfer coefficient for upward airflow conditions significantly surpassed that for downward airflow conditions [42].

For horizontal airflow, when the air velocity is $w > 0.2$ m/s, the general equation for the forced convection that outlines the relationship of the whole body on the air velocity w is given as:

$$h_c = a_f \cdot w^{b_f} \quad (16)$$

where a_f and b_f are the coefficients determined by the specific analyses carried out by the researchers.

The studies on the convective heat transfer have been carried out on the human body, or on manikins that emulate the human body.

For the studies on the human body, as discussed in [44], the relationships between the air velocity w and the convective heat transfer coefficient h_c obtained from experimental studies of frontal wind on the human body show large discrepancies among the values of a_f and b_f, which increase with the air velocity [34].

Table 2 reports the coefficients used in the representations found by different authors.

Table 2. Coefficients for the representations of the convective heat transfer coefficients of the human body under forced convection.

w (m/s)	a_f	b_f	Reference
Downward air currents	12.1	0.5	[45]
(Not indicated)	8.6	0.531	[46]
$0.10 < w < 2$	6.51	0.391	[47]
$0.15 < w < 1.5$	14.8	0.69	[48]
(Not indicated)	8.3	0.5	[49]
> 1	8.7	0.6	[50]
(Not indicated)	8.3	0.6	[51]
Still air	8.6	0.5	[46]

Both theoretical and experimental methods have been applied to heated thermal manikins with varied body shapes, sizes, and postures. A thermal manikin is a device used for evaluating human thermal environments because it accurately replicates human heat production, heat dissipation, and body shape [52]. The measurements of the heat transfer coefficients for every part of the body are impacted by the ambient factors, specifically, the temperature difference between the human body and its surroundings, the wind velocity, and the body's position and walking speed [53].

A literature review reveals that, in the context of thermal manikin experiments involving forced convection, researchers typically disregarded the influence of free convection, due to the condition Ar << 1 [54]. However, when a variable temperature gradient is present, there can be situations where the effects of natural and forced convection are comparable. In such cases, it is inappropriate to neglect either of these processes. In combined natural and forced convection scenarios, both natural and forced convection mechanisms play a role in heat transfer [54].

Over time, many studies have tried to create a useful database for predicting the convective heat transfer coefficient for the human body. A general form of the empirical formulas where the convective heat transfer coefficient is influenced by the air velocity, as seen in Equation (16). Table 3 shows the coefficients determined by different studies with the use of manikins.

Table 3. Coefficients for the representations of the convective heat transfer coefficients under forced convection by using manikins.

w (m/s)	Position	Note	a_f	b_f	Reference
$1.08 < w < 12.67$	Sitting		16.731	0.573	[55]
$0.15 < w < 0.2$	Standing		15.4	0.63	[56]
$0.2 < w < 5.5$, upstream flow	Sitting or standing	Nude	9.31	0.60	[57]
$0.2 < w < 5.5$, downstream flow	Sitting or standing	Nude	9.41	0.61	[57]
$0.2 < w < 5.5$, upstream flow	Sitting or standing	Clothed	13.36	0.60	[57]
$0.2 < w < 5.5$, downstream flow	Sitting or standing	Clothed	12.38	0.65	[57]
$0.2 < w < 0.8$	Sitting		10.1	0.61	[30]
$0.2 < w < 0.8$	Standing		10.4	0.56	[30]
$0.2 < w < 0.8$	Sitting or standing		10.3	0.6	[30]
$w < 6$	Walking		8.17	0.43	[58]
$w < 6$	Standing		7.34	0.49	[58]

There are some discrepancies in the values shown in Table 3, which can be partially explained by body geometry and body posture, as the difference for standing and sitting postures using the same manikin. Moreover, some discrepancies could be explained by the fact that many studies neglect the turbulence intensity.

When turbulence intensity is taken into account, the formulation of the convective heat transfer coefficient becomes more articulate than the structure of Equation (16). Ono et al. [59] determined the convective heat transfer coefficient for the human body in an outdoor environment using a combination of wind tunnel testing and computational fluid dynamics analysis. They proposed a new formula for the mean convective heat transfer coefficient for the human body in standing position at ambient temperature of 30 °C, as a function dependent on wind velocity w and turbulence intensity T_I:

$$h_c = 4 \cdot w + 0.035 \cdot w \cdot T_I - 8 \cdot 10^{-4} \cdot (w \cdot T_I)^3 + 3.5 \tag{17}$$

The prediction of the heat transfer coefficient considering both turbulence and wind intensity on the human body can be performed in outdoor conditions where the wind velocity is high. This formula is useful for assessing thermal comfort in an outdoor environment where there is big turbulence. Nevertheless, their proposed formula does not provide a transparent explanation of the underlying principle, and the turbulence intensities studied are lower than those commonly observed in typical pedestrian-level urban microclimates, which typically reach around 30% [60].

Yu et al. [44] measured the heat loss from the human body's surface on a thermal manikin in a simulated outdoor urban wind environment with realistic ranges of wind velocity from 0.7 m/s to 6.7 m/s and turbulence intensity from 13% to 36%. The ambient temperature is 19 °C. The regression equation for estimating convective heat transfer between the body surface and outdoor urban surroundings is expressed as:

$$h_c = a \cdot w^n \cdot \left(1 + b \cdot T_I \cdot w^{0.5}\right) \tag{18}$$

where a and n depend on the shape of the segments, and b is less affected. When T_I is set to 4%, the regression formula, Equation (18) yields results that closely align with the findings of [39], which report turbulence intensity (T_I) values ranging from 4% to 8%. However, when compared to the study in [59], the discrepancies become more pronounced with increasing wind velocity, exceeding 20% when the velocity is higher than 3 m/s. This divergence can be attributed to the fact that their experiments were conducted at wind velocities below 2 m/s.

The indications provided above refer to the whole body. However, many studies have presented results for individual body segments. De Dear et al. [30] conducted a study on an unclothed thermal manikin consisting of 16 body segments (head, chest, back, upper arms, forearms, hands, pelvis, upper legs, lower legs, and feet) generating convective and radiative heat transfer coefficients under various microclimatic conditions, both indoor and outdoor, using a climate chamber and wind tunnel. The study focused on wind speeds from calm air conditions up to 5 m/s, and eight wind directions. In addition, the thermal manikin (both seated and standing postures) remained stationary throughout the experiments. The study estimated that the radiative heat transfer coefficient was 4.7 W/(m²·K) and the convective heat transfer coefficient ranged from 3.3 to 3.4 W/(m²·K) when the air velocity was below 1 m/s and the temperature difference between the skin and the surrounding air was 12 K. The conclusions of their study were that limbs, especially hands and feet, had higher convective heat transfer coefficients than the torso. The head and neck had the lowest coefficients due to the insulative shoulder-length hair on the manikin. Seated and standing postures had similar heat losses in still air but seated showed slightly more loss in moving air. The conclusion drawn in [39] is that, under controlled airflow conditions, the convective heat transfer coefficients for the clothed manikin significantly exceeded those of the nude manikin, with disparities ranging from 100% to 200% for individual body parts and 30% to 50% for the entire body.

Luo et al. [61] investigated how movement speed, direction angle, and temperature difference between the human body and its environment affect convective heat transfer coefficients. Experiments with a thermal manikin in a cabin demonstrated that movement speed has a stronger impact on upper limbs than trunk parts. Convective heat transfer coefficients are influenced by the movement direction, with higher losses observed when moving against the wind. Additionally, coefficients increase with higher movement speeds and temperature differences.

Fojtlín et al. [62] carried out a study that aimed to experimentally determine heat transfer coefficients using a thermal manikin, with a focus on repeated coefficient measurements and statistical analysis. The manikin imitated the human metabolic heat production, measuring combined dry heat flux and surface temperature while reducing radiative heat flux with a low-emissivity coating. Tests were conducted across 34 body zones in both standing and seated postures, maintaining constant air temperature (24 °C) and wind speed (0.05 m/s). Their conclusion was that sitting manikins had slightly higher convective heat transfer coefficients compared to standing, while the opposite was observed for radiative heat transfer coefficients, consistent with other studies. Mean heat transfer coefficient values closely aligned with the existing literature, with minor variations in specific body segments. Overall, reproducibility of the measurements was achieved.

Yang and Zhang [63] analyzed how the convective heat transfer coefficient of the human body changes at different body angles (0–180°) and air velocities (0.2 to 20 m/s). The convective heat transfer coefficient values for both the entire body and specific body parts increase following a power exponent function as air velocity rises. Generally, higher air velocity led to increased heat transfer convective coefficients, with hands, feet, and limbs having higher values than the trunk except at a 90° body angle. The impact of body angle on heat transfer coefficient varied by body segment and air velocity. The following regression equation was developed in this study to express the convective heat transfer coefficient as a function of velocity and human body angle for both the entire body and individual body segments:

$$h_c = h_0 \cdot \left(z_0 + a \cdot \left(\frac{H_a}{180°} \right) + b \cdot w + c \cdot \left(\frac{H_a}{180°} \right)^2 + d \cdot w^2 \right) \tag{19}$$

where h_0 is the convective heat transfer coefficient at a human body angle of 0° and the air velocity of 0.2 m/s, H_a represents the human body angle, a, b, c, and d are dimensionless regression coefficients, and z_0 is a constant term. Equation (12) derived in the study demonstrates strong predictive accuracy for heat transfer convective coefficient at air velocities of 0.5 m/s and 4.9 m/s, but it exhibits a maximum relative error exceeding 10% at air velocities of 2 m/s and 2.8 m/s.

2.4. Radiative Heat Transfer

Radiative heat transfer of the human body refers to the process by which heat is emitted from the surface of a person's skin in the form of electromagnetic radiation, primarily in the infrared portion of the electromagnetic spectrum. This type of heat transfer occurs due to the temperature difference between the human body and its surroundings. When a person's skin is warmer than the objects or air around them, their body emits heat in the form of infrared radiation. This radiation carries energy away from the body and into the environment, helping to regulate the body's temperature and maintain thermal comfort.

The radiative thermal flux density from the clothes to the surrounding air is expressed by the following equation [30,64]:

$$\dot{q}_{rad} = f_{clo} \cdot h_r \cdot (\bar{t}_{clo} - \bar{t}_r) \qquad (20)$$

where h_r is the radiative heat transfer coefficient and $\bar{t}_r$ is the mean radiant temperature perceived by the body.

The radiative heat transfer coefficient is computed as:

$$h_r = 4 \cdot \varepsilon \cdot \sigma \cdot \left(\frac{A_r}{A_D}\right) \cdot \left[273.2 + \frac{\bar{t}_{cl} + \bar{t}_r}{2}\right]^3 \qquad (21)$$

where ε is the average body surface emissivity, $\sigma = 5.67 \cdot 10^{-8}$ W/(m²·K) is the Stefan-Boltzmann constant, A_r is the effective radiation area of the human body, A_D is the DuBois body surface area [65], and $\bar{t}_r$ is the mean radiant temperature for each segment.

The emissivity of the body can be assumed as 0.95, as already adopted by other manikin owners ([30,36]), and the effective radiation area factor Arad/ADuBois for the standing posture is generally accepted as 0.73 while for a sitting person was estimated to be 0.70 [65]. The value of radiative heat transfer coefficient is $h_r = 4.7 \frac{W}{m^2 \cdot K}$ and is suitable for general applications within standard indoor temperature conditions. The estimation of h_r values for individual body segments produced important findings, which were presented in [30,37].

2.5. Combined Convective and Radiative Heat Transfer

When investigating the exchange of heat between the human body and its surroundings, it is essential to consider the total sensible heat released through conduction, convection, and radiation. Conductive heat exchange depends on the surface area of body parts in direct contact with external surfaces, and it is neglected when there is a minor contact with solid objects [31,62]. In this case, the total sensible heat released is the sum between the convective heat exchange and radiative heat exchange, which can be calculated for clothed body and for naked body. In the case of the clothed human body, the overall sensible heat released through convection and radiation is calculated using the following equation [64]:

$$\dot{q}_s = h_c \cdot f_{clo} \cdot (\bar{t}_{clo} - t_{surr\ air}) + h_r \cdot f_{clo} \cdot (\bar{t}_{clo} - \bar{t}_r) \qquad (22)$$

In the case of the naked human body, the overall sensible heat released through convection and radiation is written as [64]:

$$\dot{q}_s = h_c \cdot (\bar{t}_{skin} - t_{surr\ air}) + h_r \cdot (\bar{t}_{skin} - \bar{t}_r) \qquad (23)$$

Kurazumi et al. [66] experimentally obtained the total sensible heat transfer coefficient (convective and radiative) using a thermal manikin in a thermal environment in which each surrounding wall temperature and air temperature were considered equal. They obtained the empirical formulas for calculating the convective heat transfer coefficient of the entire body in the case of nude seated position downdraft at the temperature of 20 °C and 26 °C:

$$h_c = 4.088 + 6.592 \cdot w^{1.715} \text{ at } 0.01 \frac{\text{m}}{\text{s}} \leq w \leq 0.73 \frac{\text{m}}{\text{s}} \text{ and } t = 20\ °C \quad (24)$$

$$h_c = 2.874 + 7.427 \cdot w^{1.345} \text{ at } 0.005 \frac{\text{m}}{\text{s}} \leq w \leq 0.71 \frac{\text{m}}{\text{s}} \text{ and } t = 26\ °C \quad (25)$$

2.6. Heat Loss through Evaporation at the Skin Surface

Evaporation relies on the process of mass transfer, during which latent heat is consumed. Heat loss through evaporation at the skin surface is a complex process that depends on a number of factors, including the evaporative heat transfer coefficient, the water vapor pressure difference between the skin and the ambient air, skin wettness, and the amount of sweat secretion.

The amount of heat lost to the environment through evaporation can be calculated using the following equation:

$$\dot{q}_e = h_e \cdot w_{skin} \cdot (p_{skin} - p_{air}) \quad (26)$$

where $\dot{q}_e$ is the evaporative heat loss in W/m^2, h_e is the evaporative heat transfer coefficient in $W/(m^2 \cdot kPa)$, w_{skin} is skin wettedness (dimensionless), p_{skin} is the water vapor pressure at the skin surface, assumed to be the pressure of saturated air at the skin temperature (kPa), and p_{air} is water vapor pressure of the ambient air in kPa.

The evaporative heat transfer coefficient is estimated from the convective heat transfer coefficient using the Lewis ratio (LR), which is about 16.5 K/kPa for indoor conditions [67]:

$$LR = \frac{h_e}{h_c} \quad (27)$$

Skin wettedness (w_{skin}) is a measure of how much of the skin is covered in sweat, which is affected by factors such as sweating and diffusion through the skin. Skin wettedness ranges from approximately 0.02 to 0.06 for normal conditions, but it increases when the human body sweats more. This is because sweat is a liquid that evaporates easily, and when it evaporates, it takes heat with it. As a result, w_{skin} is also related to warm discomfort, with values $w_{skin} > 0.2$ that is perceived as uncomfortable. This is because when the skin is too wet, the sweat cannot evaporate quickly enough, and the body cannot cool down effectively [67]. In theory, w_{skin} can attain a value of 1.0 while the body maintains thermoregulatory control. However, in practice, it is difficult to exceed 0.8 [68].

2.7. Skin Temperature Regulation

After investigating the heat transfer mechanisms of conduction, convection, radiation and evaporation, the thermal conditions experienced by the human body are significantly impacted. In a neutral environmental condition where no thermoregulatory effort is needed to keep thermal equilibrium, the skin temperature usually ranges from 30 to 34 °C. The human body can be likened to a thermostat set at approximately 37 °C. This internal thermostat generally maintains a temperature regulation accuracy of around ±0.5 °C, although the temperature of the body's outer surface (the skin) can fluctuate [69].

Different factors, such as environmental conditions like temperature, airflow rate, air pressure, and humidity, as well as clothing insulation, can impact the skin's temperature and how it is distributed across the human body [21]. The research developed in [70] highlights the dynamic nature of skin temperature regulation and the role of the neck as a prominent heat loss area in cold conditions. In their research, skin temperature distribution was studied under neutral, warm, and cold conditions. From the experiments carried out in neutral conditions, the mean skin temperature remained stable during an exposure of two hours, with fluctuations limited within 0.1 °C. In a warm environment (31.5 °C), and increased by 0.6 °C due to vasodilation, with a 2.7 °C difference between neck and calf. After 2 h in a cold environment (15.6 °C), the mean skin temperature decreased by 1 °C due to vasoconstriction, with the neck being the warmest region and a key heat loss area in cold conditions.

The skin temperature also changes during physical activity to regulate the body temperature. Infrared thermography reveals that these changes are dynamic, with initial decreases in mean skin temperature due to vasoconstriction. This reduction persists throughout exercise, especially during high-intensity activities. However, during constant-load exercises, skin temperature may increase slightly, reflecting the interaction between vasoconstriction and vasodilation as body temperature rises. Different body regions exhibit varied temperature changes, with the upper limbs experiencing more pronounced decreases (2.5 to 3.75 °C after 30 min of running) than areas near working muscles, such as the calves (approximately 1.5 °C decrease) [21,71].

While the core temperature plays a significantly more substantial role in body's thermoregulation than the skin temperature (e.g., approximately 10:1 for sweating and 4:1 for shivering), both skin and core temperatures are equally important when assessing subjective thermal sensation [72]. When the skin temperature is within a narrow range near the point of thermal neutrality (where neither hot nor cold is felt), people's perception of temperature remains unchanged [68,73,74].

The regulation of the skin temperature is closely linked to the development and application of w-TEGs for human body energy harvesting. Our body's ability to maintain skin temperature within a comfortable range is a fundamental aspect of thermoregulation. A w-TEG utilizes the temperature gradient between the skin and the surrounding environment to generate electricity. By continuously monitoring and managing skin temperature, these devices can optimize their energy harvesting efficiency without compromising the wearer's thermal comfort. This relationship between skin temperature regulation and w-TEGs highlights the potential for sustainable power sources for wearable technology, offering extended battery life while ensuring that the body remains thermally comfortable and well regulated, especially in various environmental conditions.

3. Analytical Heat Transfer Equations Related to the Interaction between Human Skin and Wearable Thermoelectric Generators

3.1. Analytical Heat Transfer Equations Related to the Human Skin

In the development of analytical heat transfer equations related to the interaction between human skin and w-TEG, the human skin is considered a quasi-homogeneous structure with three layers, in which each layer has its own thermal conductivity (as shown in Figure 2). The deepest layer is fat, followed by the dermis, and finally, the outer layer is the epidermis [75].

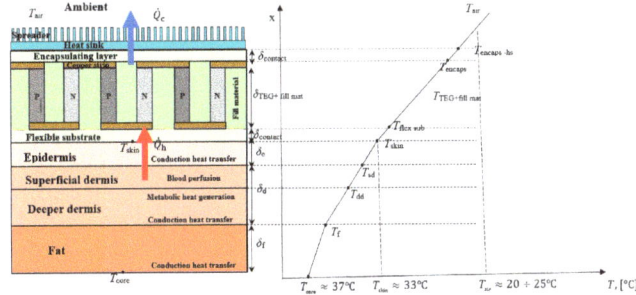

Figure 2. Sketch of the w-TEG device and the temperature distribution from the human body to the ambient (The red arrow represents the direction of the heat flux that is absorbed at the skin–w-TEG

interface. The blue arrow represents the direction of the heat flux that is released through the encapsulating layer including the spreader at the heat sink).

The general differential heat conduction equation used to describe the temperature distribution within biological tissues, including the human skin, when subject to heat sources is given by the Pennes bioheat equation [76] having the basic form as described in Equation (28):

$$\nabla^2 T + \frac{1}{k} \cdot (\dot{q}_{met} + \dot{q}_{blood}) = \frac{1}{a} \cdot \frac{\partial T}{\partial \tau} \qquad (28)$$

where $\nabla^2 T$ is the Laplacian of temperature, k is the thermal conductivity of the tissue, $\dot{q}_{met}$ is heat generation due to metabolic activity, such as cellular metabolism, $\dot{q}_{blood}$ is convective heat transfer due to blood perfusion (it considers the convective heat transfer between blood vessels and surrounding tissue), $a = \frac{k}{\rho \cdot c_p}$ is the thermal diffusivity and quantifies how quickly heat is conducted through a material compared to how quickly it is stored or accumulated within the material, ρ is the tissue density, c_p is the specific heat capacity of the tissue at constant pressure, T is the tissue temperature, and τ is the time.

The general differential heat conduction equation for the steady-state conditions within a multilayer structure with three layers, for one-dimensional geometry, is reduced to:

$$\frac{d^2 T_i}{dx^2} + \frac{1}{k_i} \cdot (\dot{q}_{met} + \dot{q}_{blood}) = 0 \qquad (29)$$

where $\frac{d^2 T_i}{dx^2}$ represents the second spatial derivative of temperature T_i with respect to the spatial coordinate T_i within of the specific layer i (such as fat, dermis and epidermis), k_i is the thermal conductivity of the specific layer i, $\dot{q}_{blood} = \rho_b \cdot c_{pb} \cdot \omega_b \cdot (T_{core} - T_i)$, ρ_b is the blood density, c_{pb} is the specific heat capacity of the blood at constant pressure, and $\omega_b = \frac{\dot{v}_b}{v_b \cdot \Delta \tau}$ is the blood perfusion rate that quantifies the rate at which blood circulates through a particular region of tissue (the blood perfusion rate can vary considerably and plays a crucial role in regulating the body's temperature), $\dot{v}_b$ is the rate of blood flow through a particular region of tissue, v_b is the volume of the tissue through which the blood is flowing, $\Delta \tau$ is the time duration over which the blood flow is being considered. When the blood flow through the skin increases, it is called vasodilation, and when it decreases it, is referred to as vasoconstriction [77]). Moreover, $\dot{v}_b$ is the tissue volume within the considered region, $\dot{v}_b$ is the volume flow rate of blood, $\Delta \tau$ is the time interval over which the blood flow is measured, T_{core} is the body core temperature, and T_i is the temperature of the skin.

In the thermal analysis of fat and epidermis layers, the influence of blood perfusion is not considered, resulting in $\dot{q}_{blood} = 0$. In this case, the influence of the blood perfusion is found only in dermis (second layer), and Equation (23) becomes [75]:

$$\frac{d^2 T_i}{dx^2} = -\frac{\dot{q}_{met}}{k_i} \qquad (30)$$

To solve the second-order derivative Equation (30) with respect to the variable x, this equation is integrated twice, taking into account the integration constants C_1 and C_2 that arise in the integration process.

For fat and epidermis, the solutions of the temperature field in each skin tissue take the form of quadratic equations:

$$T_i(x) = -\frac{\dot{q}_{met}}{2k_i} \cdot x^2 + C_{1i} \cdot x + C_{2i} \qquad (31)$$

For dermis, the solution of the temperature field is more complex and involves exponential terms:

$$T_d(x) = T_{core} - \left(\rho_b \cdot c_{pb} \cdot \omega_b\right)^{-1} \cdot \dot{q}_{met} + C_{1d} \cdot e^{m \cdot x} + C_{2d} \cdot e^{-m \cdot x} \qquad (32)$$

where $m = \left(\rho_b \cdot c_{pb} \cdot \omega_b \cdot k_d^{-1}\right)^{0.5}$.

The coefficients C_1 and C_2 for fat and epidermis as well as C_{1d} and C_{2d} for dermis are determined by applying the boundary conditions and are shown in Tables 4 and 5.

Table 4. Integration constants for fat and epidermis of the quasi-homogeneous structure.

Integration Constants	Fat i = fat	Epidermis i = epidermis
C_1	$C_{1f} = \frac{\dot{q}_{met}}{2} \cdot \frac{\delta_f}{k_f} - \delta_f^{-1} \cdot (T_{core} - T_f)$	$C_{1e} = \frac{\dot{q}_{met}}{k_e} \cdot \left(\frac{\delta_e}{2} + \delta_f + \delta_d\right) - \delta_e^{-1} \cdot (T_d - T_e)$
C_2	$C_{2f} = T_f$	$C_{2e} = -\frac{\dot{q}_{met}}{2k_e} \cdot (\delta_e + \delta_f + \delta_d) \cdot (\delta_f + \delta_d) + \frac{T_d}{2} \cdot \left(1 + \frac{\delta_f + \delta_d}{\delta_e}\right) - T_e \cdot \frac{\delta_f + \delta_d}{\delta_e}$

Table 5. Integration constants for dermis of the quasi-homogeneous structure.

Integration Constants	Dermis
C_1	$C_{1d} = \dfrac{T_d - T_f \cdot e^{-m \cdot \delta_d} - \left[T_{core} + \dot{q}_{met} \cdot (\rho_b \cdot c_{pb} \cdot \omega_b)^{-1}\right] \cdot (1 - e^{-m \cdot \delta_d})}{\left(e^{m \cdot \delta_d} - e^{-m \cdot \delta_d}\right) \cdot e^{m \cdot \delta_f}}$
C_2	$C_{2d} = \dfrac{T_d - T_f \cdot e^{m \cdot \delta_d} - \left[T_{core} + \dot{q}_{met} \cdot (\rho_b \cdot c_{pb} \cdot \omega_b)^{-1}\right] \cdot (1 - e^{m \cdot \delta_d})}{\left(e^{m \cdot \delta_d} - e^{-m \cdot \delta_d}\right) \cdot e^{-m \cdot \delta_f}}$

3.2. Wearable TEG as a Thermal Load

In a w-TEG, the extrinsic temperature gradient between the body core T_{core} and the ambient air T_{air}, $\Delta T_e = T_{core} - T_{air}$ results in a constant heat flux through the w-TEG. This heat flux leads to an intrinsic temperature gradient $\Delta T_i = T_{hot\,TEG} - T_{cold\,TEG}$ across the hot and cold sides of the w-TEG legs, which in turn produces an output voltage due to the Seebeck effect [78].

The intrinsic temperature gradient is expressed as [70]:

$$\Delta T_i = T_{hot\,TEG} - T_{cold\,TEG} = (T_{core} - T_{air}) \cdot R_{t\,TEG} \cdot (R_{skin} + R_{t\,TEG} + R_{air})^{-1} \qquad (33)$$

where $R_{t\,TEG}$ is the thermal resistance of the w-TEG, R_{skin} is the thermal contact resistance between the skin body and the w-TEG, and R_{air} is the convective thermal resistance of the surrounding air. All resistances are in K/W.

The thermal resistance of the w-TEG is given by [79]:

$$R_{t\,TEG} = \frac{1}{FF} \cdot \frac{L}{k \cdot S_{TEG}} \qquad (34)$$

where k is the thermal conductivity of the thermoelectric legs and $FF = \frac{S_N + S_P}{S_{w-TEG}}$ is the fill factor (the ratio between the area of the thermoelectric legs to the total surface area of the w-TEG).

Effectively extracting heat from the human body depends on maximizing intrinsic temperature difference. This proves to be a challenging task because other parasitic thermal resistances make it hard for the heat to flow, namely: (i) the thermal resistance between the body core and w-TEG, due to the body skin, known as a thermal insulator; (ii) the contact thermal resistance between the skin interface and w-TEG due to the skin roughness; and (iii) the convective thermal resistance at the interface between the w-TEG and the surrounding air. The convective thermal resistance at the interface between the w-TEG and surrounding air is the dominant thermal resistance if no heat sink is used at the cold side of the w-TEG [19]. However, if the w-TEG is to be used on the human body, it is favorable not to use a bulky heat sink or to use a heat sink with a small design while achieving an acceptable heatsink resistance at this interface. In this case, the heat sink must be compact, not too large or bulky, to be suitable for use on the human body without causing discomfort [78]. Enhancing the overall performance of a w-TEG is achievable through the incorporation of heat spreaders. A heat spreader offers the advantage of reduced weight and increased flexibility compared to a finned heat sink, which is particularly beneficial for wearable applications. The utilization of high thermal conductivity heat spreaders on both the cold and hot sides of a w-TEG can effectively boost the power output [19].

When a w-TEG is in contact with the skin, on the one hand the w-TEG limits the heat transfer from the body to the surrounding air, acting as a thermal barrier. On the other hand, at the same time the w-TEG absorbs heat from the skin to generate electricity, acting as a thermal load. Figure 3 shows the thermal circuit, with indicative values of the temperatures for air, skin, and core.

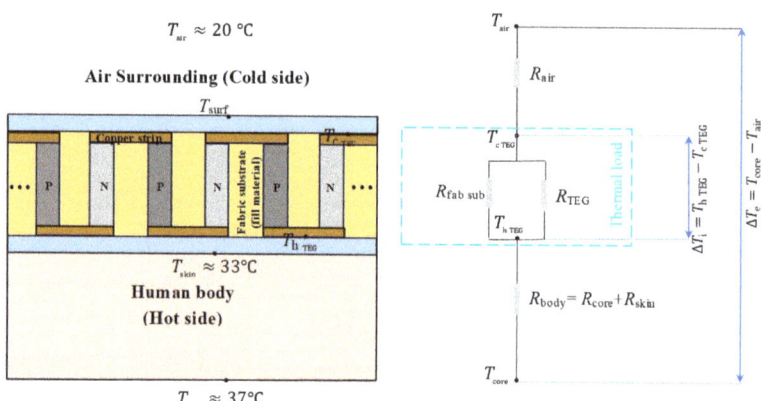

Figure 3. Thermal circuit of the w-TEG with n thermocouples used on the human body.

The simplified model of the thermal circuit is presented [70,80]. The optimal thermal resistance of the thermal circuit is defined as follows:

$$R_{\text{optimal TEG}} = \frac{1}{2 \cdot R_{\text{fab sub}}^{-1} + \left(R_{\text{body}} + R_{\text{air}}\right)^{-1}} \quad (35)$$

where $R_{\text{fab sub}}$ is the parasitic thermal resistance due to the fabric material, which is in parallel with the total thermal resistance of the w-TEG with n thermoelements; $R_{\text{body}} = R_{\text{core}} + R_{\text{skin}}$ is the thermal resistance between the body core and the skin surface; and R_{air} is the thermal resistance of the surrounding air.

Nevertheless, the thermal resistance of the thermoelectric components is variable as a result of the Peltier effect during w-TEG operation. For this reason, it is proposed to consider an effective thermal conductivity concept when calculating the thermal resistance of the w-TEG, as provided in the following equation [81]:

$$R_{\text{t TEG}} = \frac{k \cdot S}{L} \cdot \left(1 + \frac{Z\overline{T}}{1 + \sqrt{Z\overline{T}}}\right) \quad (36)$$

where k is the thermal conductivity, L is the thermoelement leg length, and S is the thermoelement area.

The thermal contact resistance R_{tc} at the interface where the w-TEG contacts the human skin has a big impact on how well the w-TEG operates. High thermal contact resistance can impede the heat flow from the skin to the w-TEG. This means that less heat is utilized for electricity generation, reducing the w-TEG conversion efficiency. Reducing thermal contact resistance through improved materials or TEG design can enhance the module performance by allowing for more efficient heat transfer, thus increasing the amount of electricity generated from the human body heat. The w-TEGs must possess flexibility to ensure they establish conformal contact and minimize thermal resistance at the skin/w-TEG interface when integrated into a self-powered wearable electronic system [75].

The resistance at the point of contact should be measured between the skin and the w-TEG surface being explored. The thermal contact resistance model utilized by Benali-Khoudja and colleagues [82] in the development of their thermal display is derived from the model originally proposed in [83]. This model considered various factors including mechanical, thermophysical, and surface properties. In the absence of any fluid in the interfacial gap, the thermal contact conductance K_{tc} at the skin and w-TEG interface can be expressed as follows [84,85]:

$$K_{tc} = \frac{1}{R_{tc}} = \underbrace{\left[1.25 \cdot k_{tc} \cdot \frac{\Delta s_{tc}}{r_{tc}} \cdot \left(\frac{p_{tc}}{\delta_{tc}}\right)^{0.95}\right]}_{h_{tc}} \cdot S_{\text{skin tiss}} \qquad (37)$$

where R_{tc} is the thermal contact resistance, h_{tc} is the heat transfer coefficient; k_s is the harmonic mean thermal conductivity for solid 1 (epidermis) and solid 2 (flexible substrate of w-TEG) with $k_{tc} = 2 \cdot \frac{k_{\text{epidermis}} \cdot k_{\text{flexible substrate of w-TEG}}}{k_{\text{epidermis}} + k_{\text{flexible substrate of w-TEG}}}$; r_{tc} is the effective root mean square roughness, with the relation $r_{tc} = \left(r^2_{\text{epidermis}} + r^2_{\text{flexible substrate of w-TEG}}\right)^{0.5}$; Δs_{tc} is effective absolute average surface asperity slope with $\Delta s_{tc} = \left(\Delta s_{\text{epidermis}}^2 + \Delta s_{\text{flexible substrate of w-TEG}}^2\right)^{0.5}$; p_{tc} is the contact pressure; δ_{tc} is the microhardness of the softer material (epidermis); and $S_{\text{skin tiss}}$ is the cross-section area of skin tissues [75].

3.3. Analytical Heat Transfer Equations Related to Wearable TEG

In order to streamline the analysis, it is assumed that both P-type and N-type legs possess symmetric material properties and dimensions. Consequently, the subsequent theoretical analysis focuses solely on the P-type thermoelement [75]. The temperature distribution in the thermoelements for the steady state is expressed as follows:

$$\frac{d^2 T_P}{dx^2} + \frac{I^2 \cdot R_P}{K_P \cdot \delta^2_{\text{TEG+fill mat}}} = 0 \qquad (38)$$

where I is the electric current that flows in the w-TEG, and $R_P = \frac{\delta_{\text{TEG+fill mat}}}{\sigma_{P \text{ eff}} \cdot S_P}$, $K_P = \frac{k_P \cdot S_P}{\delta_{\text{TEG+fill mat}}}$, $K_P = R_{tP}^{-1}$ are the electric resistance, thermal conductance, and thermal resistance of the P-type thermoelectric leg, respectively.

It is important to highlight that, in contrast to thermoelectric legs, the Joule heat generation in metallic electrodes is negligible, as presented in [75].

The electric current does not flow through the flexible layers and in this case, its value is $I = 0$. In this condition, Equation (41) becomes:

$$\frac{d^2 T_n}{dx^2} = 0 \qquad (39)$$

where "n" represents three combinations: n = f for the flexible substrate and the skin/w-TEG interface, n = N for the N-type and P-type thermoelectric legs and the fill material, and n = ES for the encapsulating layer and the spreader at the heat sink (Figure 4).

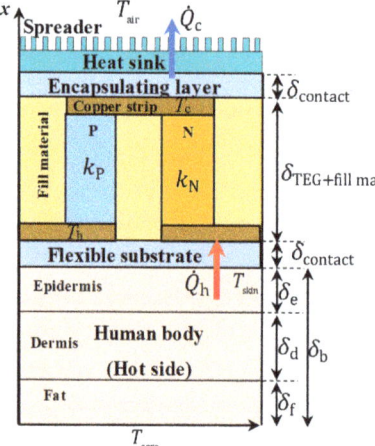

Figure 4. Schematic diagram of the w-TEG with one thermocouple (The red arrow represents the direction of the heat flux that is absorbed at the skin–w-TEG interface. The blue arrow represents the direction of the heat flux that is released through the encapsulating layer including the spreader at the heat sink).

As the thermal resistances of the thermoelectric legs and fill material are connected in parallel, the effective thermal conductivity is calculated as follows:

$$k_{ef} = k_N \cdot FF + k_{fill\ mat} \cdot (1 - FF) \tag{40}$$

where k_P is the thermal conductivity of the P-type thermoelement, $k_{fill\ mat}$ is the thermal conductivity of the fill material, and FF is the fill factor.

The temperature field distribution in the w-TEG and fill material satisfies the following equation:

$$T_P(x) = -\frac{I^2 \cdot R}{(K_N + K_P + K_{fill\ mat}) \cdot \delta_{TEG+fill\ mat}^2} \cdot x^2 + A_P \cdot x + B_P \tag{41}$$

where the constants A_P and B_P are:

$$A_P = \frac{T_c - T_h}{\delta_{TEG+fill\ mat}} + \frac{I^2 \cdot R}{(K_N + K_P + K_{fill\ mat}) \cdot \delta_{TEG+fill\ mat}^2} \cdot (\delta_b + \delta_{contact} + 0.5 \cdot \delta_{TEG+fill\ mat}) \tag{42}$$

$$B_P = T_h \cdot \left(\frac{\delta_b + \delta_{contact}}{\delta_{TEG+fill\ mat}} + 1\right) - T_c \cdot \frac{(\delta_b + \delta_{contact})}{\delta_{TEG+fill\ mat}} - \frac{I^2 \cdot R}{(K_N + K_P + K_{fill\ mat}) \cdot \delta_{TEG+fill\ mat}^2} \cdot (\delta_b + \delta_{contact}) \\ \cdot (\delta_b + \delta_{contact} + \delta_{TEG+fill\ mat}) \tag{43}$$

where $K_N = \frac{k_N \cdot S_N}{\delta_{TEG+fill\ mat}}$ is thermal conductance of the N-type thermoelectric leg, $K_P = \frac{k_P \cdot S_P}{\delta_{TEG+fill\ mat}}$ is thermal conductance of the P-type thermoelectric leg, $K_{fill\ mat} = \frac{k_{fill\ mat}}{\delta_{TEG+fill\ mat}} \cdot S \cdot (1 - FF)$ is the thermal conductance of the fill material, $R = R_N + R_P = \frac{\delta_P}{\sigma_P \cdot S_P} + \frac{\delta_P}{\sigma_N \cdot S_N}$ is the electrical resistance of the TEG thermoelements.

The heat flux that is absorbed at the skin–w-TEG interface including the flexible substrate is:

$$\dot{Q}_h = K_{equiv\ 1} \cdot (T_{skin} - T_{flex\ sub}) \tag{44}$$

where $K_{equiv\ 1} = \frac{K_{tc} \cdot K_{contact}}{K_{tc} + K_{contact}}$ is the equivalent thermal conductance between the thermal contact conductance, at the skin and w-TEG interface K_{tc} (as expressed in Equation (37)), and the thermal conductance of the flexible substrate $K_{contact}$.

The heat flux that is released through the encapsulating layer including the spreader at the heat sink is:

$$\dot{Q}_c = K_{equiv2} \cdot (T_{encap} - T_{air}) \tag{45}$$

where $K_{equiv2} = \frac{K_{encaps} \cdot K_{spreader}}{K_{encaps} + K_{spreader}}$ is the equivalent thermal conductance between the encapsulating layer $K_{encaps} = \frac{k_{encaps} \cdot S_{encaps}}{\delta_{contact}}$ and the thermal conductance of the spreader $K_{spreader} = h_{air} \cdot S_{spreader}$ at the heat sink, k_{encaps} is the thermal conductivity of the encapsulating layer, h_{air} is the convective coefficient of the air, and S_{encaps} and $S_{spreader}$ are the cross-sectional area for the encapsulating layer and for the spreader, respectively.

According to the steady-state energy conservation law, heat flow rates must remain continuous at the interfaces between various components, including fat, dermis, epidermis, substrate, thermoelectric legs (including fill material), and the encapsulating layer [75,86]. In this case: $\dot{Q}_f(\delta_f) = \dot{Q}_d(\delta_f)$; $\dot{Q}_d(\delta_f + \delta_d) = \dot{Q}_e(\delta_f + \delta_d)$; $\dot{Q}_e(\delta_b) = \dot{Q}_{substr++skin-w-TEG}(\delta_b)$; $\dot{Q}_{substr++skin-w-TEG}(\delta_b + \delta_{contact}) = \dot{Q}_{w-TEG+fill\,mat}(\delta_b + \delta_{contact})$; $\dot{Q}_{w-TEG+fill\,mat}(\delta_b + \delta_{contact} + \delta_{TEG+fill\,mat}) = \dot{Q}_{encaps+spreader}(\delta_b + \delta_{contact} + \delta_{TEG+fill\,mat})$. These conditions are necessary to obtain the expressions of the temperatures for the hot and cold sides of the w-TEG, T_h and T_c, as well as the temperatures T_e, T_d, and T_f at the interfaces of the skin tissues, as shown in Table 6 [75].

Table 6. The temperature expressions for w-TEG hot and cold sides, as well as the skin tissue interfaces.

Temperature	Temperature Expressions	Relationships Involved in Temperature Expressions
Temperature at the hot side of the w-TEG	$T_h = \dfrac{E_0 \cdot \left(\alpha \cdot I - K_{TEG+\text{fill mat}} - K_{\text{equiv 2}}\right) - K_{TEG+\text{fill mat}} \cdot \left(K_{\text{equiv 2}} \cdot T_{air} + \frac{I^2 \cdot R}{2}\right)}{K_{TEG+\text{fill mat}}^2 + E_1 \cdot \left(\alpha \cdot I - K_{TEG+\text{fill mat}} - K_{\text{equiv 2}}\right)}$	$E_0 = D_0 + \dfrac{I^2 \cdot R}{2}$ $E_1 = D_1 + \alpha \cdot I + K_{\text{equiv 1}} + K_{TEG+\text{fill mat}}$ $D_0 = -\big[K_{\text{equiv 1}} \cdot \delta_e \cdot \dot{Q}_{met} \cdot (D_e \cdot D_f - D_d^2)$ $+ K_{\text{equiv 1}} \cdot K_e \cdot K_d \cdot D_d \cdot T_{substr} + K_{\text{equiv 1}} \cdot K_e \cdot \dot{Q}_{met} \cdot$ $(\delta_e \cdot D_f + \delta_f \cdot D_d) - K_{\text{equiv 1}} \cdot K_e \cdot D_d \cdot R \cdot (D_d + D_f)\big] \cdot$ $\dfrac{1}{\big[K_e^2 \cdot D_f - (K_{\text{equiv 1}} + K_e)(D_e \cdot D_f - D_d^2)\big]}$ $D_1 = \dfrac{K_h^2 \cdot (D_e \cdot D_f - D_d^2)}{\big[K_e^2 \cdot D_f - (K_{\text{equiv 1}} + K_e)(D_e \cdot D_f - D_d^2)\big]}$
Temperature at the cold side of the w-TEG	$T_c = -\dfrac{E_0 \cdot K_{TEG+\text{fill mat}} + E_1 \cdot \left(K_{\text{equiv 2}} \cdot T_{air} + \frac{I^2 \cdot R}{2}\right)}{K_{TEG+\text{fill mat}}^2 + E_1 \cdot (\alpha \cdot I - K_{TEG+\text{fill mat}} - K_{\text{equiv 2}})}$	
Temperature on the top surface of epidermis	$T_e = \dfrac{(K_{\text{equiv 1}} \cdot T_h + \delta_e \cdot \dot{Q}_{met}) \cdot (D_e \cdot D_f - D_d^2) + K_e \cdot K_d \cdot D_d \cdot T_{substr} + K_e \cdot \dot{Q}_{met} \cdot (\delta_e \cdot D_f + \delta_f \cdot D_d) - K_e \cdot D_d \cdot P \cdot (D_d + D_1)}{K_e^2 \cdot D_f - (K_{\text{equiv 1}} + K_e) \cdot (D_e \cdot D_f - D_d^2)}$	$D_d = \dfrac{K_d \cdot m \cdot \delta_d}{\sinh(m \cdot \delta_d)}$ $D_e = K_e + D_d \cdot \cosh(m \cdot \delta_d)$ $D_f = K_f + D_d \cdot \cosh(m \cdot \delta_d)$ $P = \left(T_{substr} + \dfrac{q_{met}}{\rho_b \cdot c_{pb} \cdot \omega_b}\right) \cdot [1 - \cosh(m \cdot \delta_d)]$ $m = \left(\rho_b \cdot c_{pb} \cdot \omega_b \cdot k_d^{-1}\right)^{0.5}$ from Equation (29)
Temperature on the top surface of dermis	$T_d = \dfrac{K_e \cdot D_f \cdot T_e + K_f \cdot D_f \cdot T_{substr} + \dfrac{\dot{Q}_{met} \cdot (\delta_e \cdot D_f + \delta_f \cdot D_d)}{D_e \cdot D_f - D_d^2} - D_d \cdot P \cdot (D_d + D_f)}{}$	
Temperature on the top surface of fat	$T_f = \dfrac{K_{\text{equiv 2}} \cdot T_{substr} + \dot{Q}_{met} \cdot \delta_f + D_d (T_s - P)}{D_f}$	

3.4. Evaluation of the w-TEG Performance

The performance analysis of w-TEG involves a detailed examination of the power output generated using the temperature differential between the heat source and heat sink [75,86,87]. The power output is calculated due to the principle of energy conservation and is evaluated as:

$$P_{out} = \dot{Q}|_{x=\delta_b+\delta_{contact}} - \dot{Q}|_{x=\delta_b+\delta_{contact}+\delta_{TEG+fill\ mat}} = \alpha \cdot I \cdot (T_h - T_c) - \frac{R \cdot I^2}{2} \quad (46)$$

where α is the Seebeck coefficient.

When an external load resistance R_L is introduced, the power output can be written as follows:

$$P_{out} = R_L \cdot I^2 \quad (47)$$

The expression for the electric current is obtained by setting Equation (46) equal to Equation (47):

$$\alpha \cdot I \cdot (T_h - T_c) - \frac{R \cdot I^2}{2} = R_L \cdot I^2 \Rightarrow I = \frac{\alpha \cdot (T_h - T_c)}{R + R_L} \quad (48)$$

The substitution of the expressions of the temperatures T_h and T_c indicated in Table 6 in Equation (48) results in a third-degree equation with the electric current as the unknown [75,86,87]:

$$a_3 \cdot I^3 + a_2 \cdot I^2 + a_1 \cdot I + a_0 = 0 \quad (49)$$

where the parameters a_3, a_2, a_1, and a_0 are known and have the following expressions [75]:

$$a_0 = \alpha \cdot K_{equiv2} \cdot [D_0 - (K_{equiv1} + D_1) \cdot T_{air}] \quad (50)$$

$$a_1 = -\left[(K_{TEG+fill\ mat} + K_{equiv2}) \cdot (K_{equiv1} + D_1) + K_{TEG+fill\ mat} \cdot K_{equiv2}\right] \cdot (R + R_L) - \alpha^2 \cdot (D_0 + K_{equiv2} \cdot T_{air}) \quad (51)$$

$$a_2 = \alpha \cdot (K_{equiv\ 1} - K_{equiv\ 2} + D_1) \cdot \left[R_L + \frac{R}{2}\right] \quad (52)$$

$$a_3 = \alpha^2 \cdot R_L \quad (53)$$

The solution of Equations (49) and (52) has the following form:

$$I = -\frac{1}{3} \cdot a_2 \cdot a_3^{-1} + \xi^2 \cdot \sqrt[3]{-\frac{\psi}{2} + \sqrt{\left(\frac{\psi}{2}\right)^2 + \left(\frac{\chi}{3}\right)^3}} + \xi \cdot \sqrt[3]{-\frac{\psi}{2} - \sqrt{\left(\frac{\psi}{2}\right)^2 + \left(\frac{\chi}{3}\right)^3}} \quad (54)$$

with $\xi = -\frac{1}{2} + i \cdot \frac{\sqrt{3}}{2}$, $\psi = \frac{a_0}{a_3} + \frac{2}{27} \cdot \left(\frac{a_2}{a_3}\right)^3 - \frac{a_1 \cdot a_2}{3 \cdot a_3^2}$, and $\chi = \frac{a_1}{a_3} - \frac{1}{3} \cdot \left(\frac{a_2}{a_3}\right)^2$.

The power output is calculated by replacing Equation (54) into Equation (47). In addition, the power density P_D of the w-TEG refers to the amount of electrical power produced by the w-TEG per unit area. The power density is the ratio between the power output P_{out} generated by the w-TEG to the surface area S over which the TEG is distributed or attached:

$$P_D = \frac{P_{out}}{S} \quad (55)$$

The performance of a w-TEG is influenced by various parameters, including external load resistance R_L, human body skin, thermoelectric couple height, and fill factor FF. Zhang et al. [87] investigated the influence of these parameters on the performance of w-TEG. The research conducted in [87] highlights the significance of these parameters in optimizing the performance of w-TEG devices. Understanding how factors such as the thermophysical properties of the filler material, human body skin, fill factor, thermal convective boundary conditions, and the figure of merit of thermoelectric legs (ZT) influence the external load resistance can lead to the development of more efficient and practical w-TEG systems, with potential benefits in energy harvesting, waste heat recovery, and sustainable power generation. The metabolic heat generated has very little effect on the power density of the skin-w-TEG system, even when the metabolic heat generation increases significantly. Conversely, the blood perfusion rate within the skin tissue plays a key role in enhancing the w-TEG performance, with power density increasing significantly as the blood perfusion rate rises. Therefore, effective management and optimization of the w-TEG should consider blood perfusion as a crucial factor. The influence of convective heat transfer between the w-TEG's lateral surfaces and the ambient surroundings significantly impacts the temperature drop (ΔT) and power density of the w-TEG. The influence of convective heat transfer on a w-TEG's lateral surfaces is essential because it determines how effectively the w-TEG can maintain the necessary temperature gradient for efficient electricity generation. If convective heat transfer is not considered or optimized, it can lead to reduced performance and power output of the w-TEG. Also, the influence of convective heat transfer at the heat sink and contact pressure between human skin and w-TEG significantly affects temperature drop and power density. Increasing contact pressure causes the human body's skin to deform. Since the human body's skin is generally softer than the flexible substrate of the w-TEG, this deformation can effectively increase the actual contact area with the substrate. This, in turn, reduces the contact thermal resistance, leading to more effective heat transfer. Concerning the influence of the height of the thermoelectric couples and the convective heat transfer coefficient, these significantly affect the temperature drop and power density of the w-TEG. Increasing the thermoelectric couple height improves the temperature drop but also raises the internal electric resistance, leading to an optimal height for maximum power density. Additionally, heat convection at lateral surfaces becomes more significant with higher thermoelectric couple heights.

In summary, the performance analysis of w-TEG involves evaluating the power output based on temperature gradients. Various factors, including external load resistance, human body skin, thermoelectric couple height, and fill factor, significantly impact the w-TEG performance. Understanding these parameters is crucial for optimizing w-TEG systems. Additionally, the influence of metabolic heat generation and blood perfusion within the skin tissue should be considered. Convective heat transfer between w-TEG lateral surfaces and the surroundings also affects temperature and power density. Optimizing contact pressure and managing thermocouple height are key factors for maximizing power density. These insights are essential for enhancing w-TEG design and applications.

4. Wearable Thermoelectric Devices for Personal Thermal Management

Personal thermal management refers to the establishment of a satisfactory thermal environment around the human body, in which the individuals live and operate under acceptable thermal comfort conditions. The solutions for personal thermal management can be used to reduce the energy needed to heat or cool the internal environments in the buildings, concentrating the means for reaching satisfactory temperatures to the space close to the activity of the individuals [88].

There are various possibilities for impacting on the local thermal environment. The first category includes non-wearable solutions, which act on the forms of heat transfer that have an impact on the human body. These solutions (whose detailed analysis is outside the scope of this article) include, for example:

- Thermally controlled chairs: thermoelectric devices are used in thermoelectrically heated and cooled chairs to have an influence on thermal sensation and comfort, as shown, for example, from the experiments presented in [89,90]. The thermal sensation can be improved when the temperature is outside the acceptable range, even though the effects of chair heating can be limited by the fact that the thermal sensation of the extremities cannot be improved to a significant extent [91].
- Systems for local heating, ventilation, and air conditioning: portable solutions have been developed for heating or cooling the local environment and interact with the thermoregulation of the human body [92]. For example, a thermoelectric air condi-

tioning undergarment solution that provides personal heating or personal cooling depending on the control mode with air volume control is illustrated in [93]. The system contains a power-supplied micro-blower that heats up or cools down the air in the local ambient and uses a system with small tubes to send the air to various parts of the human body.

The other category, which is of interest for this section, is the one of wearable solutions, consisting of thermoregulated clothing of different types. Traditional thermoelectric devices are generally based on inorganic thermoelectric devices—which are rigid, have poor mechanical properties, and are expensive—so that they can be applied to localized solutions (e.g., sensors) more than for clothing purposes [94]. For the diffusion of wearable thermoelectric solutions for clothing, the main aspect is the development of flexible devices, which can adapt to the movements of the human body [95] or can be easily stretched [96] and are composed of sections that are simply reconfigurable [97]. In particular, flexible thermoelectric devices based on organic composites, such as polymer materials [94], are interesting because of their specific properties of flexibility, low toxicity, low cost, and stability at relatively high temperatures, even though their efficiency is lower than the that obtained with inorganic materials [98]. The development of organic–inorganic composite thermoelectric devices and other hybrid solutions [99] is promising, with the possibility of benefiting from the best properties of the different materials and providing cost-effective applications [100]. In the evolution in progress on the materials side, promising solutions are expected from film-based thermoelectric modules, because of significant performance, high scalability, and opportunities for largescale production [101].

For the development of wearable solutions, the expressions indicated in the previous sections illustrate the basic heat transfer principles referring to the interactions between the human body and the wearable TEG. In particular, the skin thermal resistance is higher than the TEG resistance and plays a considerable role in wearable applications. Further practical aspects depend on body comfort. Generally, wearable solutions are acceptable if their size and weight are limited. For this purpose, the preferred solutions have lower thermal conductivity of the material that forms the thermoelectric legs, in such a way that a thinner TEG can be constructed by maintaining a relatively large temperature difference.

Moreover, in some practical applications, the use of a heat sink is avoided because of esthetic or space reasons. However, without a heat sink, the power density that can be obtained from a wearable TEG is generally limited with respect to the power density obtained from a wearable TEG with heat sink. For wearable TEG applications, solutions that use a flat thermal spreader to replace the heat sink can be more adaptable. Even though in general the thermal spreader could be less efficient than a heat sink for dissipating heat, the w-TEG presented in [8] with a flat thermal spreader reached a 30% power density increase with respect to the best reported TEG with heat sinks.

Active and passive thermal management methods have been categorized [102] for wearable solutions referring to heating or cooling modes, as follows:

- Active methods: in general, thermoelectric devices can be used for cooling and heating, as well as electro- and magnetocaloric cooling and heating. For active heating, the typical source is Joule heating, and for active cooling the active microfluidic cooling is adopted. Regarding active heating, thermoelectric textiles based on the Peltier effect, in which flexible thermoelectric devices are integrated into the textiles to provide power generation, can be more efficient than Joule heating textiles [103]. For active cooling with thermoelectric devices, typically, the circulation of water in a copper tube is added for improving the heat exchange; the cooling output that can be provided is relatively low, due to the low coefficient of performance, and could be enhanced with the use of multistage thermoelectric modules [104]. An effective solution for a wearable solution with a thermoelectric device that does not use a water heat sink and can produce a cooling effect of more than 10 °C by maintaining a relatively high coefficient of performance is presented in [105].

- Passive methods: in general, heat storage from the external environment is obtained with materials having high latent heat or high heat capacity to store and release heat as needed. Further methods include thermal insulation to minimize the heat transfer with respect to the human skin. For passive heating purposes, thermally conductive materials are used to enhance the heat exchange with the air, or photothermal materials are used to absorb solar energy to warm the human skin. For passive cooling purposes, radiative cooling materials are used to refrigerate the human skin, and evaporative cooling materials facilitate the transition from liquid to vapor. Passive methods are not based on thermoelectric devices.

The wearable solutions can be partitioned, for example, into:

- Natural heat exchange: these solutions focus on natural heat exchange and energy harvesting and aim to harness the body's natural heat production and the surrounding environment to maintain thermal comfort.
- Assisted heat exchange: these solutions adopt clothing enhancements and include additional parts with fans for better air circulation, or in other cases water circulation systems for making the temperatures in the different parts of the clothing more uniform. The effectiveness of these solutions could depend on the type of activity carried out by the individual in the living environment. For example, the extra devices that allow for assisted heat exchange could add weight or size to the clothing, potentially reducing the mobility of the individuals when carrying out certain activities.

From another point of view, the devices used in the wearable solutions can be categorized into:

- Autonomous devices: self-powered solutions in which there is no energy input from external sources. These solutions rely on internal energy sources to regulate temperature and maintain thermal comfort.
- Non-autonomous devices: solutions for which an energy input is needed from external sources. These solutions require a continuous supply of energy to function and regulate temperature effectively.
- Hybrid devices: These combine nonautonomous and autonomous devices in different parts of the wearable solutions. Some parts of the clothing may operate autonomously while others rely on external energy sources. This hybrid approach offers flexibility in managing thermal comfort. The review presented in [106] addresses many cases of personal comfort devices and indicates an energy efficient solution with combined use of air-cooling units and a thermoelectric cooling unit with limited surface coverage.

A further distinction refers to the type of contact of the thermoelectric devices with the human body, which is an important aspect that also affects the choice of the materials and of the modes for connecting the devices to the human body:

- For direct contact with the human body, the main solutions include cooling vests with the thermoelectric device in contact with human skin [105]. A key aspect is to avoid the contact of the human body with rare or toxic elements that can be found in some thermoelectric devices (e.g., bismuth, lead, or tellurium) [107]. Biobased thermoelectric materials (such as cotton, cellulose, or lignin), which have less impact on the human body, can be used as a substrate for constructing wearable devices.
- Without direct contact with the human body (i.e., with indirect contact), the heat transfer modes have to be studied by considering the materials used for clothing. The use of flexible and long thermoelectric fibers is an effective solution for covering the various possible curvatures of the surfaces, enhancing thermal management and comfort [108].

In summary, personal thermal management is a multifaceted field with a wide range of solutions for enhancing thermal comfort and energy efficiency. These solutions can range from nonwearable solutions that affect heat transfer around the human body to wearable solutions, including thermoregulated clothing. The advancement in materials and technology has enabled the development of flexible and efficient thermoelectric devices that can be integrated into clothing, offering active and passive thermal management methods. Furthermore, the choice between autonomous, nonautonomous, or hybrid devices, as well as the consideration of direct or indirect contact with the human body, provides a wide array of options for tailoring personal thermal management solutions to individual needs and preferences. The exploration of these possibilities holds promise for improving comfort and wellbeing of individuals, while reducing energy consumption in various environments.

5. Conclusions

Technological development in the thermoelectric generation area is providing interesting solutions to create better local thermal environments for individuals who carry out activities in indoor environments. This paper addressed wearable thermoelectric generators that interact with the human body to improve local thermal comfort and increase energy efficiency by reducing, to a certain extent, the need for heating or cooling the ambient. The analysis of the heat transfer mechanisms that appear in the interaction between thermoelectric devices and the human body was discussed as a crucial step for understanding how the human thermoregulatory system can respond in the presence of wearable thermoelectric generators that consider the body as a heat source. Moreover, the review presented in this paper summarized how conductive, convective, and radiative heat transfer, together with evaporation at the skin surface, impact on the temperature regulation of the skin, which is the surface of contact between the human body and thermoelectric devices. Furthermore, the materials, flexibility, and efficiency of wearable thermoregulated clothing that can be used to create a comfortable thermal environment around the human body were considered, also categorizing their active and passive modes of operation.

Wearable thermoelectric generators are versatile and adaptable solutions to address individual thermal needs in local environments. The main directions for the future are the development of thermoelectric materials with enhanced thermal performance and the design of structurally flexible, lightweight, and cost-effective solutions able to enhance the interaction with the human body considering situations with parts of the body in movement. On the materials side, biobased materials and organic composites are of interest because of their low environmental impact. Thermal performance refers to the efficiency of the energy conversion, for which the use of multistage thermoelectric modules and the exploitation of smart sensors and control systems can provide more adaptive thermal regulation.

The evolution of wearable thermoelectric generation technologies and application requires multidisciplinary collaborations between experts of different domains, such as materials science, thermal engineering, environmental engineering, physiology, psychology, up to clothes design, for merging their competences in the direction of providing more comfortable, energy efficient, cost-effective, and practically appealing solutions.

Funding: This research received no external funding.

Data Availability Statement: No new data were created or analyzed in this study.

Conflicts of Interest: The author declares no conflicts of interest.

References

1. Hasan, M.N.; Sahlan, S.; Osman, K.; Ali, M.S.M. Energy harvesters for wearable electronics and biomedical devices. *Adv. Mater. Technol.* **2021**, *6*, 2000771. [CrossRef]
2. Enescu, D. Book Chapter "Thermoelectric energy harvesting: Basic Principles and Applications". In *Green Energy Advances*; Enescu, D., Ed.; InTech Publishing: Rijeka, Croatia, 2019; pp. 1–37.
3. Du, Y.; Xu, J.; Paul, B.; Eklund, P. Flexible thermoelectric materials and devices. *Appl. Mater. Today* **2018**, *12*, 366–388. [CrossRef]
4. He, J.; Li, K.; Jia, L.; Zhu, Y.; Zhang, H.; Linghu, J. Advances in the applications of thermoelectric generators. *Appl. Therm. Eng.* **2024**, *236*, 121813. [CrossRef]
5. Siddique, A.R.M.; Mahmud, S.; Heyst, B.V. A review of the state of the science on wearable thermoelectric power generators (TEGs) and their existing challenges. *Renew. Sustain. Energy Rev.* **2017**, *73*, 730–744. [CrossRef]
6. Sahin, A.Z.; Yilbas Bekir, S. The thermoelement as thermoelectric power generator: Effect of leg geometry on the efficiency and power generation. *Energy Convers Manag.* **2013**, *65*, 26–32. [CrossRef]

7. Francioso, L.; De Pascali, C.; Farella, I.; Martucci, C.; Cretì, P.; Siciliano, P.; Perrone, A. Flexible thermoelectric generator for ambient assisted living wearable biometric sensors. *J. Power Sources* **2011**, *196*, 3239–3243. [CrossRef]
8. Nozariasbmarz, A.; Suarez, F.; Dycus, J.H.; Cabral, M.J.; LeBeau, J.M.; Öztürk, M.C.; Vashaee, D. Thermoelectric generators for wearable body heat harvesting: Material and device concurrent optimization. *Nano Energy* **2020**, *67*, 104265. [CrossRef]
9. Shi, Y.; Wang, Y.; Mei, D.; Chen, Z. Wearable Thermoelectric Generator with Copper Foam as the Heat Sink for Body Heat Harvesting. *IEEE Access* **2018**, *6*, 43602–43611. [CrossRef]
10. Zoui, M.A.; Bentouba, S.; Stocholm, J.G.; Bourouis, M. A Review on Thermoelectric Generators: Progress and Applications. *Energies* **2020**, *13*, 3606. [CrossRef]
11. Yuan, J.; Zhu, R.; Li, G. Self-powered electronic skin with multisensory functions based on thermoelectric conversion. *Adv. Mater. Technol.* **2020**, *5*, 000419. [CrossRef]
12. Kim, J.; Khan, S.; Wu, P.; Park, S.; Park, H.; Yu, C.; Kim, W. Self-charging wearables for continuous health monitoring. *Nano Energy* **2021**, *79*, 105419. [CrossRef]
13. Hasan, M.N.; Asri, M.I.A.; Saleh, T.; Muthalif, A.G.A.; Ali, M.S.M. Wearable thermoelectric generator with vertically aligned PEDOT:PSS and carbon nanotubes thermoelements for energy harvesting. *Int. J. Energy Res.* **2022**, *46*, 15824–15836. [CrossRef]
14. Kim, C.S.; Yang, H.M.; Lee, J.; Lee, G.S.; Choi, H.; Kim, Y.J.; Lim, S.H.; Cho, S.H.; Cho, B.J. Self-powered wearable electrocardiography using a wearable thermoelectric power generator. *ACS Energy Lett.* **2018**, *3*, 501–507. [CrossRef]
15. Lee, B.B.; Cho, H.; Park, K.T.; Kim, J.-S.; Park, M.; Kim, H.; Hong, Y.; Chung, S. High-performance compliant thermoelectric generators with magnetically self-assembled soft heat conductors for self-powered wearable electronics. *Nat. Commun.* **2020**, *11*, 1. [CrossRef]
16. Kong, D.; Zhu, W.; Guo, Z.; Deng, Y. High-performance flexible Bi2Te3 films based wearable thermoelectric generator for energy harvesting. *Energy* **2019**, *175*, 292–299. [CrossRef]
17. Khan, S.; Kim, J.; Roh, K.; Park, G.; Kim, W. High power density of radiative-cooled compact thermoelectric generator based on body heat harvesting. *Nano Energy* **2021**, *87*, 106180. [CrossRef]
18. Choi, H.; Kim, Y.J.; Song, J.; Kim, C.S.; Lee, G.S.; Kim, S.; Park, J.; Yim, S.H.; Park, S.H.; Hwang, H.R.; et al. UV-curable silver electrode for screen-printed thermoelectric generator. *Adv. Funct. Mater.* **2019**, *29*, 1901505. [CrossRef]
19. Nozariasbmarz, A.; Collins, H.; Dsouza, K.; Polash, M.H.; Hosseini, M.; Hyland, M.; Liu, J.; Malhotra, A.; Ortiz, F.M.; Mohaddes, F.; et al. Review of wearable thermoelectric energy harvesting: From body temperature to electronic systems. *Appl. Energy* **2020**, *258*, 114069. [CrossRef]
20. Nandihalli, N.; Liu, C.-J.; Mori, T. Polymer based thermoelectric nanocomposite materials and devices: Fabrication and characteristics. *Nano Energy* **2020**, *78*, 105186. [CrossRef]
21. Francioso, L.; De Pascali, C. Chapter 10 Thermoelectric Energy Harvesting for Powering Wearable Electronics. In *Thermoelectric Energy Conversion: Basic Concepts and Device Applications*, 1st ed.; Dávila Pineda, D., Alireza Rezania, A., Eds.; Wiley-VCH: Weinheim, Germany, 2017; pp. 205–231.
22. Kim, G.H.; Shao, L.; Zhang, K.; Pipe, K.P. Engineered doping of organic semiconductors for enhanced thermoelectric efficiency. *Nat. Mater.* **2013**, *12*, 719. [CrossRef]
23. de Fazio, R.; Cafagna, D.; Marcuccio, G.; Minerba, A.; Visconti, P. A Multi-Source Harvesting System Applied to Sensor-Based Smart Garments for Monitoring Workers' Bio-Physical Parameters in Harsh Environments. *Energies* **2020**, *13*, 2161. [CrossRef]
24. Tanwar, A.; Lal, S.; Razeeb, K. Structural Design Optimization of Micro-Thermoelectric Generator for Wearable Biomedical Devices. *Energies* **2021**, *14*, 2339. [CrossRef]
25. Enescu, D. Models and Indicators to Assess Thermal Sensation Under Steady-state and Transient Conditions. *Energies* **2019**, *12*, 841. [CrossRef]
26. Lai, D.T.H.; Palaniswami, M.; Begg, R. *Healthcare Sensor Networks: Challenges toward Practical Implementation*; CRC Press, Taylor & Francis Group: Boca Raton, FL, USA, 2016; 462p.
27. Merriam-Webster Dictionary, "Thermal Inertia". Available online: https://www.merriam-webster.com/dictionary/thermal%20inertia (accessed on 28 November 2023).
28. Houdas, Y.; Ring, E.F.J. *Human Body Temperature: Its Measurement and Regulation*; Springer Science & Business Media: New York, NY, USA, 2013; 238p.
29. Kuwabara, K.; Mochida, T.; Nagano, K.; Shimakura, K. Experiments to determine the convective heat transfer coefficient of a thermal manikin. *Elsevier Ergon. Book Series* **2005**, *3*, 423–429.
30. de Dear, R.J.; Arens, E.; Hui, Z.; Oguro, M. Convective and radiative heat transfer coefficients for individual human body segments. *Int. J. Biometeorol.* **1997**, *40*, 141–156. [CrossRef] [PubMed]
31. Oliveira, A.V.; Gaspar, A.R.; Francisco, S.C.; Quintela, D.A. Convective heat transfer from a nude body under calm conditions: Assessment of the effects of walking with a thermal manikin. *Int. J. Biometeorol.* **2012**, *56*, 319–332. [CrossRef] [PubMed]
32. Cena, K.; Clark, J.A. *Bioengineering, Thermal Physiology, and Comfort*; Elsevier: Amsterdam, The Netherlands, 1981.
33. Psikuta, A.; Kuklane, K.; Bogdan, A.; Havenith, G.; Annaheim, S.; Rossi, R.M. Opportunities and constraints of presently used thermal manikins for thermo-physiological simulation of the human body. *Int. J. Biometeorol.* **2015**, *60*, 435–446. [CrossRef]
34. Xu, J.; Psikuta, A.; Li, J.; Annaheim, S.; Rossi, R.M. Influence of human body geometry, posture and the surrounding environment on body heat loss based on a validated numerical model. *Build. Environ.* **2019**, *166*, 106340. [CrossRef]
35. Fanger, P.O. *Thermal Comfort*; Danish Technical Press: Copenhagen, Denmark, 1970.

36. Quintela, D.; Gaspar, A.; Borges, C. Analysis of sensible heat exchanges from a thermal manikin. *Eur. J. Appl. Physiol.* **2004**, *92*, 663–668. [CrossRef]
37. Kurazumi, Y.; Tsuchikawab, T.; Ishiia, J.; Fukagawaa, K.; Yamatoc, Y.; Matsubarad, N. Radiative and convective heat transfer coefficients of the human body in natural convection. *Build. Environ.* **2008**, *43*, 2142–2153. [CrossRef]
38. Nielsen, M.; Pedersen, L. Studies on the heat loss by radiation and convection from the clothed human body. *Acta Physiol. Scand.* **1953**, *27*, 272–294. [CrossRef] [PubMed]
39. Oguro, M.; Arens, E.; de Dear, R.; Zhang, H.; Katayama, T. Convective heat transfer coefficients and clothing insulations for parts of the clothed human body under calm conditions. *J. Archit. Plan. Environ. Eng.* **2002**, *561*, 31–39.
40. Omori, T.; Yang, J.H.; Kato, S.; Murakami, S. Coupled simulation of convection and radiation on thermal environment around an accurately shaped human body. In Proceedings of the RoomVent 2004, 9th International Conference on Air Distribution in Rooms, Coimbra, Portugal, 5–8 September 2004.
41. Sparrow, E.M.; Eichhorn, R.; Gregg, J.L. Combined forced and free convection in a boundary layer flow. *Phys. Fluids* **1959**, *2*, 319. [CrossRef]
42. Gao, S.; Ooka, R.; Oh, W. Formulation of human body heat transfer coefficient under various ambient temperature, air speed and direction based on experiments and CFD. *Build. Environ.* **2019**, *160*, 106168. [CrossRef]
43. Colin, J.; Houdas, Y. Experimental determination of coefficient of heat exchanges by convection of human body. *J. Appl. Physiol.* **1967**, *22*, 31–38. [CrossRef] [PubMed]
44. Yu, Y.; Liu, J.; Chauhan, K.; de Dear, R.; Niu, J. Experimental study on convective heat transfer coefficients for the human body exposed to turbulent wind conditions. *Build. Environ.* **2020**, *169*, 106533. [CrossRef]
45. Winslow, C.-E.A.; Gagge, A.P.; Herrington, L.P. The influence of air movement on heat losses from the clothed human body. *J. Physiol.* **1939**, *127*, 505–515. [CrossRef]
46. Gagge, A.P.; Fobelets, A.P.; Berglund, L.G. A standard predictive index of human response to the thermal environment. *ASHRAE Trans.* **1986**, *92*, 709–731.
47. Nishi, Y.; Gagge, A.P. Direct evaluation of convective heat transfer coefficient by naphthalene sublimation. *J. Appl. Physiol.* **1970**, *29*, 830–838. [CrossRef]
48. Seppänen, O.; McNall, P.E.; Munson, D.M.; Sprague, C.H. Thermal insulating values for typical indoor clothing ensembles. *ASHRAE Trans.* **1972**, *78*, 120–130.
49. Kerslake, D.M.K. *The Stress of Hot Environments*; Cambridge University Press: Cambridge, UK, 1972.
50. Missenard, F.A. Coefficients d'échange de chaleur du corps humain per convection, en function de la position, de l'activité du sujet et de l'environment. (In French): Heat Exchange Coefficients of the Human Body by Convection, as a Function of Position, Subject's Activity, and Environment. *Arh. Sci. Physiol.* **1973**, *27*, A45–A50.
51. Mitchell, D. Convective heat transfer from man and other animals. In *Heat Loss from Animals and Man*; Monteith, J.L., Mount, L.E., Eds.; Elsevier: London, UK, 1974; pp. 59–76.
52. Fukazawa, T.; Tochihara, Y. The thermal manikin: A useful and effective device for evaluating human thermal environments. *J. Human-Environ. Syst.* **2015**, *18*, 21–28. [CrossRef]
53. Fu, M.; Weng, W.; Chen, W.; Luo, N. Review on modeling heat transfer and thermoregulatory responses in human body. *J. Therm. Biol.* **2016**, *62 Pt B*, 189–200. [CrossRef]
54. Luo, N.; Weng, W.G.; Fu, M. Theoretical analysis of the effects of human movement on the combined free-forced convection. *Int. J. Heat Mass Transf.* **2015**, *91*, 37–44. [CrossRef]
55. Li, C.; Ito, K. Numerical and experimental estimation of convective heat transfer coefficient of human body under strong forced convective flow. *J. Wind Eng. Ind. Aerod.* **2014**, *126*, 107–117. [CrossRef]
56. Ichihara, M.; Saitou, M.; Tanabe, S.; Nishimura, M. Measurement of convective heat transfer coefficient and radiative heat transfer coefficient of standing human body by using thermal manikin. In Proceedings of the Annual Meeting of the Architectural Institute of Japan, Tokyo, Japan, August 1995; pp. 379–380.
57. Oguro, M.; Arens, E.; de Dear, R.; Zhang, H.; Katayama, T. Convective heat transfer coefficients and clothing insulations for parts of the clothed human body under airflow conditions. *J. Archit. Plan. Environ. Eng. AIJ* **2002**, *561*, 21–29.
58. Oliveira, A.V.M.; Gaspar, A.R.; Francisco, S.C.; Quintela, D.A. Analysis of natural and forced convection heat losses from a thermal manikin: Comparative assessment of the static and dynamic postures. *J. Wind Eng. Ind. Aerod.* **2014**, *132*, 66–76. [CrossRef]
59. Ono, T.; Murakami, S.; Ooka, R.; Omori, T. Numerical and experimental study on convective heat transfer of the human body in the outdoor environment. *J. Wind Eng. Ind. Aerodyn* **2008**, *96*, 1719–1732. [CrossRef]
60. Roth, M. Review of atmospheric turbulence over cities. *Q. J. R. Meteorol. Soc.* **2000**, *126*, 941–990. [CrossRef]
61. Luo, N.; Weng, W.G.; Fu, M.; Yang, J.; Han, Z.Y. Experimental study of the effects of human movement on the convective heat transfer coefficient. *Exp. Therm. Fluid Sci.* **2014**, *57*, 40–56. [CrossRef]
62. Fojtlín, M.; Fišer, J.; Jícha, M. Determination of convective and radiative heat transfer coefficients using 34-zones thermal manikin: Uncertainty and reproducibility evaluation. *Exp. Therm. Fluid Sci.* **2016**, *77*, 257–264. [CrossRef]
63. Yang, J.; Zhang, S. Three-dimensional simulation of the convective heat transfer coefficient of the human body under various air velocities and human body angles. *Int. J. Therm. Sci.* **2023**, *187*, 108171. [CrossRef]
64. Fanger, P.O. *Thermal Comfort—Analysis and Applications in Environmental Engineering*; McGraw-Hill: New York, NY, USA, 1972.

65. Du Bois, D.; Du Bois, E.F. Clinical calorimetry: Tenth paper a formula to estimate the approximate surface area if height and weight be known. *Arch. Intern. Med.* **1916**, *17*, 863–871. [CrossRef]
66. Kurazumi, Y.; Rezgals, L.; Melikov, A.K. Convective heat transfer coefficients of the human body under forced convection from ceiling. *J. Ergonom.* **2014**, *4*, 1000126. [CrossRef]
67. Arens, E.; Zhang, H. The Skin's Role in Human Thermoregulation and Comfort. In *Thermal and Moisture Transport in Fibrous Materials*; Pan, N., Gibson, P., Eds.; Woodhead Publishing Ltd.: Cambridge, UK, 2006; pp. 560–602.
68. Berglund, L.G.; Gonzalez, R.R. Evaporation of sweat from sedentary man in humid environments. *J. Appl. Physiol.* **1977**, *42*, 767–772. [CrossRef] [PubMed]
69. Leonov, V.; Vullers, R.J.M. Wearable thermoelectric generators for body-powered devices. *J. Electron. Mater.* **2009**, *38*, 1491–1498. [CrossRef]
70. Huizenga, C.; Zhang, H.; Arens, E.; Wang, D. Skin and core temperature response to partial and whole-body heating and cooling. *J. Therm. Biol.* **2004**, *29*, 549–558. [CrossRef]
71. Tanda, G. The Use of Infrared Thermography to Detect the Skin Temperature Response to Physical Activity. 33rd UIT (Italian Union of Thermo-fluid-dynamics) Heat Transfer Conference. *J. Phys. Conf. Ser.* **2015**, *655*, 012062. [CrossRef]
72. Frank, S.M.; Raja, S.N.; Bulcao, C.F.; Goldstein, D.S. Relative contribution of core and cutaneous temperatures to thermal comfort and autonomic responses in humans. *J. Appl. Physiol.* **1999**, *86*, 1588–1593. [CrossRef]
73. Zhang, H. Human Thermal Sensation and Comfort in Transient and Non-Uniform Thermal Environments. Ph.D. Thesis, CEDR, University of California at Berkeley, Berkeley, CA, USA, 2003.
74. McIntyre, D.A. *Indoor Climate*; Applied Science Publishers: London, UK, 1980.
75. Zhang, A.; Li, G.; Wang, B.; Wang, J. A Theoretical Model for Thermoelectric Generators Considering the Effect of Human Skin. *J. Electron. Mater.* **2021**, *50*, 1514–1526. [CrossRef]
76. Pennes, H.H. Analysis of Tissue and Arterial Blood Temperatures in the Resting Human Forearm. *J. Appl. Physiol.* **1948**, *1*, 93. [CrossRef] [PubMed]
77. Wijethunge, D.; Kim, D.; Kim, W. Simplified human thermoregulatory model for designing wearable thermoelectric devices. *J. Phys. D Appl. Phys.* **2018**, *51*, 055401. [CrossRef]
78. Suarez, F.; Nozariasbmarz, A.; Vashaee, D.; Öztürk, M.C. Designing thermoelectric generators for self-powered wearable electronics. *Energy Environ. Sci.* **2016**, *9*, 2099–2113. [CrossRef]
79. Yee, S.K.; LeBlanc, S.; Goodson, K.E.; Dames, C. $ per W metrics for thermoelectric power generation: Beyond ZT. *Energy Environ. Sci.* **2013**, *6*, 2561–2571. [CrossRef]
80. Leonov, V. Thermoelectric energy harvester on the heated human machine. *J. Micromech. Microeng.* **2011**, *21*, 125013. [CrossRef]
81. Baranowski, L.L.; Snyder, G.J.; Toberer, E.S. Effective thermal conductivity in thermoelectric materials. *J. Appl. Phys.* **2013**, *113*, 204904. [CrossRef]
82. Benali-Khoudja, M.; Hafez, M.; Alexandre, J.M.; Benachour, J.; Kheddar, A. Thermal Feedback Model for Virtual Reality. In Proceedings of the International Symposium on Micromechatronics and Human Science, Nagoya, Japan, 19–22 October 2003; pp. 153–158.
83. Yovanovich, M.M.; Antonetti, V.W. Application of Thermal Contact Resistance Theory to Electronic Packages. In *Advances in Thermal Modeling of Electronic Components and Systems*; Bar-Cohen, A., Kraus, A.D., Eds.; Hemisphere Publishing: New York, NY, USA, 1988.
84. Ho, H.N.; Jones, L.A. Modeling the Thermal Responses of the Skin Surface During Hand-Object Interactions. *J. Biomech. Eng.-T ASME* **2008**, *130*, 021005. [CrossRef]
85. Bejan, A.; Kraus, A.D. *Heat Transfer Handbook*; John Wiley & Sons, Inc.: Hoboken, NJ, USA, 2003.
86. Pang, D.; Zhang, A.; Guo, Y.; Wu, J. Energy harvesting analysis of wearable thermoelectric generators integrated with human skin. *Energy* **2023**, *282*, 128850. [CrossRef]
87. Zhang, A.; Pang, D.; Lou, J.; Wang, J.; Huang, W.M. An analytical model for wearable thermoelectric generators harvesting body heat: An opportunistic approach. *Appl. Therm. Eng.* **2024**, *236*, 121658. [CrossRef]
88. Hu, R.; Liu, Y.; Shin, S.; Huang, S.; Ren, X.; Shu, W.; Cheng, J.; Tao, G.; Xu, W.; Chen, R.; et al. Emerging Materials and Strategies for Personal Thermal Management. *Adv. Energy Mater.* **2020**, *10*, 1903921. [CrossRef]
89. Pasut, W.; Zhang, H.; Arens, E.; Kaam, S.; Zhai, Y. Effect of a heated and cooled office chair on thermal comfort. *HVAC&R Res.* **2013**, *19*, 574–583.
90. Deng, Q.; Wang, R.; Li, Y.; Miao, Y.; Zhao, J. Human thermal sensation and comfort in a non-uniform environment with personalized heating. *Sci. Total Environ.* **2017**, *578*, 242–248. [CrossRef] [PubMed]
91. Yang, H.; Cao, B.; Zhu, Y. Study on the effects of chair heating in cold indoor environments from the perspective of local thermal sensation. *Energy Build.* **2018**, *180*, 16–28. [CrossRef]
92. Zhao, D.; Lu, X.; Fan, T.; Wu, Y.S.; Lou, L.; Wang, Q.; Fan, J.; Yang, R. Personal thermal management using portable thermoelectrics for potential building energy saving. *Appl. Energy* **2018**, *218*, 282. [CrossRef]
93. Lou, L.; Shou, D.; Park, H.; Zhao, D.; Wu, Y.S.; Hui, X.; Yang, R.; Kan, E.K.; Fan, J. Thermoelectric air conditioning undergarment for personal thermal management and HVAC energy saving. *Energy Build.* **2020**, *226*, 110374. [CrossRef]
94. Xu, S.; Shi, X.L.; Dargusch, M.; Di, C.; Zou, J.; Chen, Z.G. Conducting polymer-based flexible thermoelectric materials and devices: From mechanisms to applications. *Prog. Mater. Sci.* **2021**, *121*, 100840. [CrossRef]

95. Lee, J.A.; Aliev, A.E.; Bykova, J.S.; de Andrade, M.J.; Kim, D.; Sim, H.J.; Lepro, X.; Zakhidov, A.A.; Lee, J.B.; Spinks, G.M.; et al. Woven-Yarn Thermoelectric Textiles. *Adv. Mater.* **2016**, *28*, 5038. [CrossRef]
96. Zhu, S.; Fan, Z.; Feng, B.; Shi, R.; Jiang, Z.; Peng, Y.; Gao, J.; Miao, L.; Koumoto, K. Review on Wearable Thermoelectric Generators: From Devices to Applications. *Energies* **2022**, *15*, 3375. [CrossRef]
97. Ren, W.; Sun, Y.; Zhao, D.; Aili, A.; Zhang, S.; Shi, C.; Zhang, J.; Geng, H.; Zhang, J.; Zhang, L.; et al. High-performance wearable thermoelectric generator with self-healing, recycling, and Lego-like reconfiguring capabilities. *Sci. Adv.* **2021**, *7*, eabe0586. [CrossRef]
98. Wang, Y.; Yang, L.; Shi, X.L.; Shi, X.; Chen, L.; Dargusch, M.S.; Zou, J.; Chen, Z.G. Flexible Thermoelectric Materials and Generators: Challenges and Innovations. *Adv. Mater.* **2019**, *31*, 1807916. [CrossRef]
99. Prunet, G.; Pawula, F.; Fleury, G.; Cloutet, E.; Robinson, A.J.; Hadziioannou, G.; Pakdel, A. A review on conductive polymers and their hybrids for flexible and wearable thermoelectric applications. *Mater. Today Phys.* **2021**, *18*, 100402. [CrossRef]
100. Jiang, Q.; Yang, J.; Hing, P.; Ye, H. Recent advances, design guidelines, and prospects of flexible organic/inorganic thermoelectric composites. *Mater. Adv.* **2020**, *1*, 1038. [CrossRef]
101. Zhou, W.; Fan, Q.; Zhang, Q.; Cai, L.; Li, K.; Gu, X.; Yang, F.; Zhang, N.; Wang, Y.; Liu, H.; et al. High-performance and compact-designed flexible thermoelectric modules enabled by a reticulate carbon nanotube architecture. *Nat. Commun.* **2017**, *8*, 14886. [CrossRef] [PubMed]
102. Jung, Y.; Kim, M.; Kim, T.; Ahn, J.; Lee, J.; Ko, S.H. Functional Materials and Innovative Strategies for Wearable Thermal Management Applications. *Nano-Micro Lett.* **2023**, *15*, 160. [CrossRef] [PubMed]
103. Farooq, A.S.; Zhang, P. Fundamentals, materials and strategies for personal thermal management by next-generation textiles. *Compos. Part A Appl. Sci. Manuf.* **2021**, *142*, 106249. [CrossRef]
104. Riffat, S.B.; Ma, X. Improving the coefficient of performance of thermoelectric cooling systems: A review. *Int. J. Energy Res.* **2004**, *28*, 753–768. [CrossRef]
105. Hong, S.; Gu, Y.; Seo, J.K.; Wang, J.; Liu, P.; Meng, Y.S.; Xu, S.; Chen, R.K. Wearable thermoelectrics for personalized thermoregulation. *Sci. Adv.* **2019**, *5*, 0536. [CrossRef] [PubMed]
106. Song, W.; Zhang, Z.; Chen, Z.; Wang, F.; Yang, B. Thermal comfort and energy performance of personal comfort systems (PCS): A systematic review and meta-analysis. *Energy Build.* **2022**, *256*, 111747. [CrossRef]
107. Zhang, L.; Shi, X.L.; Yang, Y.L.; Chen, Z.G. Flexible thermoelectric materials and devices: From materials to applications. *Mater. Today* **2021**, *46*, 62–108. [CrossRef]
108. Zhang, T.; Li, K.; Zhang, J.; Chen, M.; Wang, Z.; Ma, S.; Zhang, N.; Wei, L. High-performance, flexible, and ultralong crystalline thermoelectric fibers. *Nano Energy* **2017**, *41*, 35–42. [CrossRef]

Disclaimer/Publisher's Note: The statements, opinions and data contained in all publications are solely those of the individual author(s) and contributor(s) and not of MDPI and/or the editor(s). MDPI and/or the editor(s) disclaim responsibility for any injury to people or property resulting from any ideas, methods, instructions or products referred to in the content.

Article

Evaluation of Performance and Power Consumption of a Thermoelectric Module-Based Personal Cooling System—A Case Study

Anna Dąbrowska [1,*], Monika Kobus [1], Łukasz Starzak [2] and Bartosz Pękosławski [2]

1. Department of Personal Protective Equipment, Central Institute for Labour Protection—National Research Institute, Wierzbowa 48, 90-133 Lodz, Poland; mokob@ciop.lodz.pl
2. Department of Microelectronics and Computer Science, Lodz University of Technology, Wólczańska 221/223 B18, 90-924 Lodz, Poland; lukasz.starzak@p.lodz.pl (Ł.S.); bartosz.pekoslawski@p.lodz.pl (B.P.)
* Correspondence: andab@ciop.lodz.pl

Abstract: Thermoelectric (TE) technology is promising for reducing thermal discomfort of workers during their routine professional activities. In this manuscript, a preliminary evaluation of a newly developed personal cooling system (PCS) with flexible TE modules is presented based on an analysis of cooling efficiency and power consumption. For this purpose, tests with human participation were performed involving the monitoring of local skin temperature changes and electrical parameters of the controller. Thanks to TE cooling, a significant reduction of local skin temperature was observed at the beginning of the experiment, reaching as much as 6 °C. However, the observed effect systematically became weaker with time, with the temperature difference decreasing to about 3 °C. Cooling efficiency stayed at the same level over the ambient temperature range from 25 °C to 35 °C. The obtained results showed that a proper fitting of the PCS to the human body is a crucial factor influencing the PCS cooling efficiency.

Keywords: Peltier effect; thermoelectric module; thermoelectric cooler; personal cooling; smart clothing

Citation: Dąbrowska, A.; Kobus, M.; Starzak, Ł.; Pękosławski, B. Evaluation of Performance and Power Consumption of a Thermoelectric Module-Based Personal Cooling System—A Case Study. *Energies* **2023**, *16*, 4699. https://doi.org/10.3390/en16124699

Academic Editor: Diana Enescu

Received: 9 May 2023
Revised: 7 June 2023
Accepted: 8 June 2023
Published: 14 June 2023

Copyright: © 2023 by the authors. Licensee MDPI, Basel, Switzerland. This article is an open access article distributed under the terms and conditions of the Creative Commons Attribution (CC BY) license (https://creativecommons.org/licenses/by/4.0/).

1. Introduction

According to the general principles of prevention described in the Council Directive 89/391/EEC [1], risks at the workplace should be avoided. However, there are many situations in which the only possible method of protecting the worker is by providing personal protective equipment. This may result from the characteristics of the technological process at the workplace, workers' mobility requirements or environmental parameters. One of the hazards that workers may be exposed to is a hot microclimate. It can cause just thermal discomfort to the worker, but it can also pose a risk to their health, leading to, e.g., a heat stroke. Heat stress can disrupt the employee's cognitive processes, which also has a negative impact on work performance. This issue is well known, among others, in the metal industry or mining. Considering the increasing ambient temperatures resulting from climate changes, people also working outside (such as in the construction sector) have started to complain about the thermal load, especially in the summertime during increased physical activity. Such professions usually require workers' mobility, therefore individual cooling systems may be a good solution to this problem.

Ventilation systems, liquid cooling, phase change materials (PCMs) and evaporative cooling are commonly used in personal cooling solutions developed so far [2]. Air cooling is usually based on the cooling of the small air space around the body with fans [2]. Another solution, using liquid CO_2 atmospheric discharges to create a cool microclimate under clothing, was proposed by Sayed et al. [3]. Liquid cooling, on the other hand, involves distributing a cooled liquid around the body by a network of tubes located in the garment [4]. There has also been some research already conducted that combined two cooling methods in one system. Lu et al. [5] developed a personal cooling system in which PCMs were supported by means of ventilation fans. They increased cooling efficiency of the proposed solution thanks to an increase in the evaporative cooling. Similarly, Qiao et al. [6] worked on the addition of PCMs into the condenser of an evaporative personal cooling system. These authors proved that the developed system is able to provide a cooling efficiency as high as 160 W for 4.5 h thanks to the provided active thermal storage by means of PCMs' addition. In spite of this advantage, the main drawbacks of the above-mentioned systems are their weight, size or limited operating times.

Thermoelectric (TE) modules can potentially help in overcoming these disadvantages. The flow of current through a module transports heat from one side to the other, thus lowering the temperature of the former. The electric power supplied to the module, which translates into the amount of the transported heat, can be easily adjusted by varying the current, as TE modules are largely resistive from the electrical point of view [7]. At present, technological research is focused on developing TE modules of high flexibility, which is very important in personal cooling systems. Despite many studies conducted on the development of efficient cooling systems using TE modules, there are few reports on testing these solutions with human participation, although such tests are very important for determining the usefulness of a particular solution.

Wei et al. [8] developed a soft-covered wearable TE device for body-heat harvesting and on-skin cooling in the form of TE pillars sandwiched between a fabric and a pin-fin soft cover [8]. They investigated the cooling efficiency with the participation of a human in a temperature-controlled chamber where the human placed their lower arm to which the developed device was attached. The tests were carried out in a stationary state, at various ambient temperatures. During the tests, the ambient temperature and skin temperature under the device were measured. The test lasted 30 min, of which the cooling lasted continuously for 20 min, and for the first and last 5 min of the test, the temperature was monitored with the device turned off. They showed that their solution allowed to reduce skin temperature by 1.5 °C [8].

A similar study of a developed flexible TE module was conducted by Hong et al. [9]. In a temperature-controlled chamber, the skin temperature was measured using a thermocouple placed under the device mounted on the arm. The tests were carried out at various ambient temperatures ranging from 22 °C to 36 °C for 10 min, with the device turned on after 5 min. The authors also conducted tests in realistic conditions while sitting indoors, walking indoors at 5 km/h and walking outdoors at 5 km/h [9].

Li et al. [10] tested a cooling vest solution under simulated conditions of use with the participation of a human. Their solution uses liquid cooling together with TE modules. In the study, temperature was measured at six cooled points during three activities, sitting, walking at a speed of 1.2 m/s and walking at a speed of 1.8 m/s, performed in an experimental cabin at a temperature of approximately 40 °C.

These works proved the potential of TE modules in supporting human thermoregulation. Their results, however, do not relate to a work environment. In this study, it was decided to carry out a comprehensive iterative process of development and testing of a personal cooling system intended to be used at the workplace, with the aim of providing a means of protection from the effects of heat. The novelty of our solution based on thermoelectric cooling with flexible TE modules and heat sinks was already demonstrated in [11], where we also highlighted the importance of the adopted TE module control method as well as their location on the human body. The aim of the work presented in this paper was to preliminarily evaluate the potential of flexible TE modules in the reduction of thermal discomfort during increased physical activity and under high ambient temperature. Electric power consumption and electronic controller efficiency were also taken into account and assessed under simulated utility conditions.

2. Materials and Methods

2.1. Tested Object

For the purpose of this study, a model of an active personal cooling system (PCS) based on TE modules was developed. The developed system consisted of seven TE modules, seven heat sinks, two temperature sensors, a power source and a dedicated electronic controller to regulate temperature through the electric power supplied to the TE modules. The TE modules were embedded into frames with Velcro that enabled their integration with custom-made clothing. Details related to the developed personal cooling system have been described in [11]. A view of the PCS with two TE modules and one heat sink mounted on the clothing showing the integration method is presented in Figure 1. In the utility tests, all seven TE modules and heat sinks were applied.

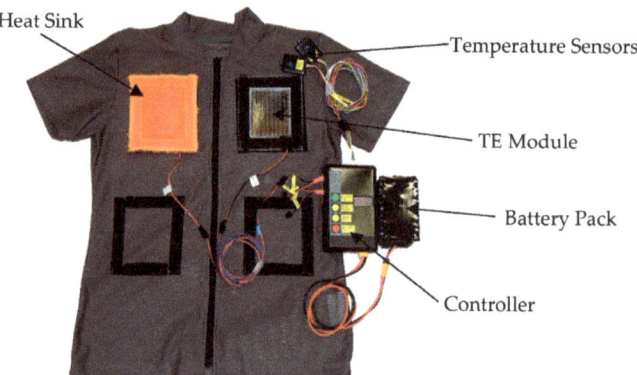

Figure 1. View of the tested system presenting TE module and heat sink integration with clothing.

The dedicated power and temperature controller is an electronic system whose block diagram is shown in Figure 2. It consists of four main functional blocks: a power processing block, an input power measurement block, an output power measurement block and a microprocessor control block. The power source for the entire system is a rechargeable battery pack.

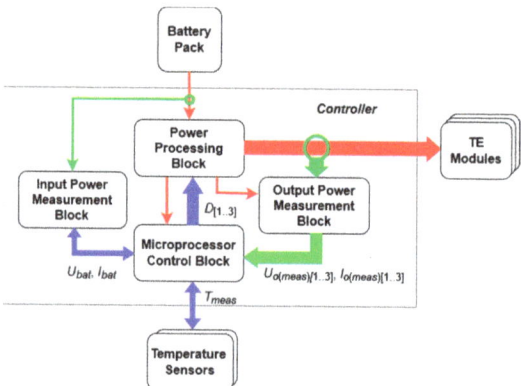

Figure 2. Block diagram of the dedicated power and temperature controller (digital signals—blue, analog signals—green, power paths—red).

The power processing block includes analog-controlled DC/DC input switched-mode power converters (SMPCs) and digitally controlled output SMPCs. The input SMPCs provide appropriate supply voltages for all the controller blocks. The output SMPCs supply the TE modules. There are three output channels, each having its own, independently controlled output SMPC.

The input power measurement block includes a battery monitor circuit that measures the battery voltage and current and converts these results into a digital form. The task of the output power measurement block is to provide analog voltage signals proportional to the instantaneous voltage and current in each output channel.

A microcontroller in the microprocessor control block reads temperature data from temperature sensors and battery voltage and current from the input power measurement block via a serial digital interface. Moreover, the microcontroller's internal analog-to-digital converter is used for converting analog measurement signals (output voltages and currents) to a digital form. The microcontroller's program includes a control algorithm enabling the digital regulation of power or temperature, as described in Section 2.2. The output signals from the microcontroller are digital pulse width-modulated (PWM) signals whose duty cycles $D_{[1..3]}$ are controlled independently.

The microprocessor control block is also equipped with a user interface (a four-button keyboard for increasing or decreasing cooling intensity and navigating through a user menu and a display for presenting system settings and status) as well as an SD card interface for storing measurement data and controller configuration in a nonvolatile memory.

2.2. Digital Controller for Output SMPCs

The output power converters are automatically controlled so that the TE modules will provide the cooling intensity level set by the user. The system can operate in one of two modes, power regulation or temperature regulation, which are also selected by the user.

In the power regulation mode, the cooling intensity corresponds to the electric supply power of a single TE module $P_{pm(set)}$. This setting is common for all the output converters and is used to determine the output power set point,

$$P_{o(set)} = N_{pm} P_{pm(set)}, \qquad (1)$$

where N_{pm} is the number of TE modules supplied by the given converter. This set point is achieved using a digital proportional controller with feedforward implemented in the microcontroller. Its structure is presented in Figure 3.

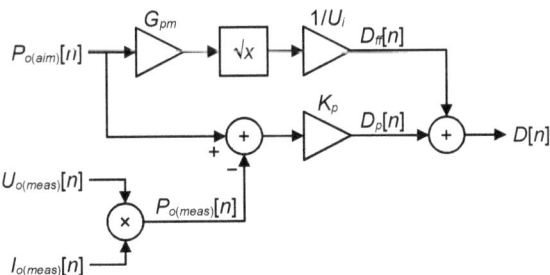

Figure 3. Block diagram of the digital power controller.

The feedforward term is calculated from the ideal buck converter equation using the conductivity of the supplied TE modules G_{pm} and the input voltage of the output converters U_i, as described in detail in [11]:

$$D_{ff} = (P_{o(aim)} G_{pm})^{1/2} / U_i, \qquad (2)$$

where $P_{o(aim)}$ is the aimed output power. In this equation, U_i is a constant parameter, as the output converters are supplied from one of the input SMPCs that delivers a regulated voltage of a constant value at its output. The conductivity G_{pm} is measured at the output of each converter when the controller is turned on. This is performed by briefly applying a low voltage and measuring the resulting current.

On the other hand, the feedback (proportional) term is calculated as

$$D_p = K_p P_{o(e)}, \qquad (3)$$

where K_p is the proportional term coefficient and $P_{o(e)}$ is the output power error,

$$P_{o(e)} = P_{o(aim)} - P_{o(meas)}. \qquad (4)$$

The actual output power $P_{o(meas)}$ is calculated based on the measurement results of the given converter's output voltage $U_{o(meas)}$ and output current $I_{o(meas)}$:

$$P_{o(meas)} = U_{o(meas)} I_{o(meas)}. \qquad (5)$$

It must be noted that the aimed output power $P_{o(aim)}$ appearing in Equations (2) and (4) is not identical with the set output power $P_{o(set)}$. The aimed power is gradually changed up or down in constant steps until a new set power is reached. In this way, abrupt strikes of cold or warmth are avoided when $P_{o(set)}$ increases or decreases. On the other hand, the power should not change too slowly as this would make the user feel as if the system underperformed. The optimum stepping rate has been found empirically to be 400 ms.

The output of the power controller represents the duty cycle to be applied to its transistor switch. It is simply the sum of the feedforward and feedback terms:

$$D = D_{ff} + D_p, \qquad (6)$$

In the temperature regulation mode, the power controller still operates according to the same equations. However, the set output power $P_{o(set)}$, to which the input signal $P_{o(aim)}$ is stepped, is no longer directly coupled to the cooling intensity setting. Instead, it is derived by a higher-level temperature controller whose block diagram is shown in Figure 4.

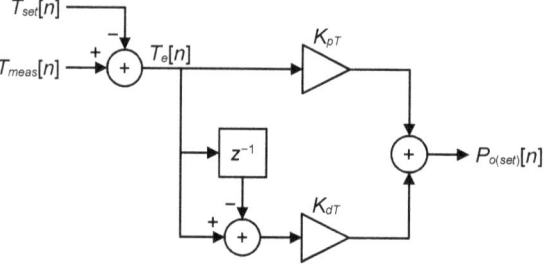

Figure 4. Block diagram of the digital temperature controller.

PID controllers are often used for temperature control using TE modules, such as in box coolers [12]. However, as argued in [11], the proportional-derivative (PD) type is optimal for the considered application due to its following features:

- Simplicity.
- Inherent stability.
- Ease of empirical term coefficient determination when plant parameters are unknown and variable (e.g., between users).
- High reactivity to fast changes of the power setting or the temperature measured.
- The corresponding controller equation is:

$$P_{o(set)} = K_{pT} T_e[n] + K_{dT} (T_e[n] - T_e[n-1]), \qquad (7)$$

where K_{pT} and K_{dT} are the proportional and derivative term coefficients, respectively, n is the current sample number and T_e is the temperature error. The latter signal is calculated according to:

$$T_e = T_{meas} - T_{set}, \qquad (8)$$

where T_{meas} is the temperature measured (an average of the readings from the two sensors) and T_{set} is the desired temperature set by the user.

It should be noted that although it is the set temperature that is subtracted from the measured one, the controller still implements a negative feedback scheme. This follows from the fact that the reaction of the input signal (temperature) is principally opposite to the change in the output signal (TE module supply power): when the electric supply power increases, heat is removed from the body more intensively, which makes the temperature decrease.

2.3. Testing Methodology

2.3.1. Research Conditions

This research was carried out at the Research and Demonstration Laboratory at the Central Institute for Labour Protection—National Research Institute, with the participation of a human tester. All tests were performed with the same participant so that results from different steps of the iterative testing and development process may be compared with each other. Tests were performed on different days at similar times. The participant was acquainted with the research methodology and signed an informed consent before starting the tests. The tests were carried out in a controlled environment where the temperature varied from 25 °C to 35 °C in steps of 5 °C. The relative humidity was constant at 65% and the air velocity was constant at 0 m/s.

2.3.2. Measured Parameters

During the experiments, thermal and electrical parameters were measured. Local skin temperature of the user was measured using AC1913-A sensors from Rotronic in two locations on the front of the body and on the back, with two sensors per location: one under a TE cooler and one beside the cooler. For the input (battery pack) voltage and current, the temperature was measured by either electronic sensor as well as the average of the two temperatures, the output voltage, the current and power of each TE module channel, the total output power of all TE module channels, the cooling level set by the user and the corresponding temperature or power.

2.3.3. Test Variants

An earlier research presented in [11] concentrated on the selection of the optimal system configuration in terms of ensuring high cooling efficiency while limiting the electric energy consumption. The following were demonstrated:

- The best results were obtained using a control mode where the TE module supply power was alternatingly regulated by the controller described in Section 2.2, with a limit of 2 W, and brought down to a standby value of 1 W, with either phase duration of 1 min;
- The optimum arrangement of TE modules was that with two modules on the front and five modules on the back.

These findings were applied in the present work, where laboratory tests were divided into two stages, as presented in Table 1. Stage I was aimed at the investigation of the effect of the active cooling using TE modules. The aim of stage II was to determine the influence of the ambient temperature. In stage I of the tests, two TE modules in the front were mounted on the upper chest, while in stage II of the tests, they were mounted on the abdomen, as preferred on the basis of the results from our previous research [9].

Table 1. Study stages and related experiment conditions.

Stage	Study	Heat Sinks	TE Modules	Ambient Temperature
I	1	None	Off	30 °C
	2	Mounted	Off	30 °C
	3	Mounted	Active	30 °C
II	4	Mounted	Active	25 °C
	5	Mounted	Active	30 °C
	6	Mounted	Active	35 °C

2.3.4. Testing Procedure

The clothing and the cooling system were acclimated in a separate air-conditioned room at a temperature of 23 °C for 1 h before the experiment in order to standardize study conditions. The participant stayed under the same conditions prior to the study. The conditions in the laboratory room were stabilized for 2 h.

Before starting the experiment, heat sinks were prepared by soaking them in water at a room temperature of 23 °C until their weight reached between 80 g and 90 g. At the same time, the TE modules were placed on the clothing in designated locations (depending on the stage of the study), which was then put on by the participant.

After soaking the heat sinks, they were attached to the TE modules using Velcro tapes. In order to achieve good adhesion of the TE modules to the skin, which is crucial for an efficient operation of the cooling system, the adherence of the clothing with PCS to the participant's body was improved by means of an additional fastening system. During the experiment, the electronic controller together with the battery pack were placed in a kidney bag at the waist of the participant.

The tests were performed using a Zebris FDM-THM-M-3i inclinable treadmill by Zebris Medical GmbH. Each of the tests at stages I and II was composed of six phases, as shown in Table 2.

Table 2. Test phases at stages I and II.

Phase	Activity	Movement Kind	Movement Speed	Track Inclination
a	I	Walk	3 km/h	0%
b	Break	None	N/A	N/A
c	II	Walk	5 km/h	0%
d	Break	None	N/A	N/A
e	III	Walk	5 km/h	10%
f	Break	None	N/A	N/A

3. Results

3.1. Effect of Active TE Modules on Cooling Efficiency (Stage I)

To assess the cooling efficiency of the developed system, local skin temperatures underneath and next to the TE modules were analyzed for the three cooling variants indicated in Table 1. Figures 5 and 6 show the temperatures recorded on the chest (Figures 5a and 6a) and on the shoulder blades (Figures 5b and 6b).

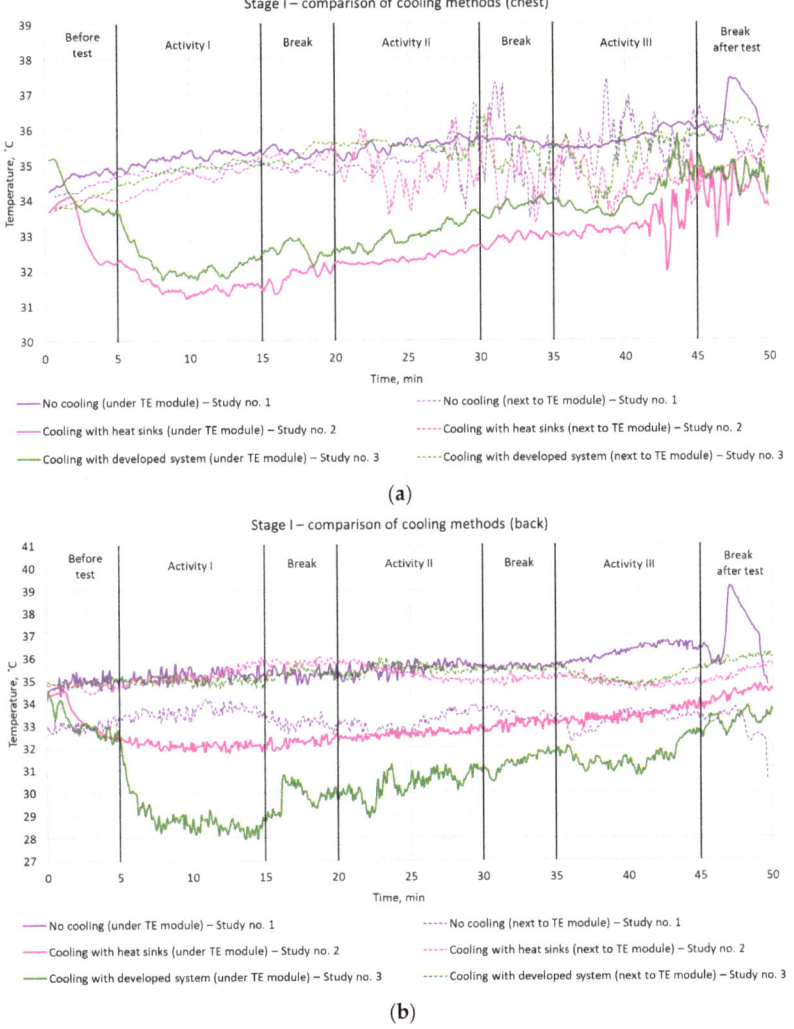

Figure 5. Local skin temperature under and next to a TE module for different cooling variants (stage I, study numbering compliant with Table 1): (**a**) chest; (**b**) back.

A significant temperature drop before starting the test was observed for both Studies 2 and 3, resulting from the soaked heat sinks being mounted. However, the effect of activating the TE modules in the PCS was much more pronounced, as observed at the beginning of Activity I in Study 3.

For both cooling variants (Studies 2 and 3), lower local skin temperatures were observed throughout an entire experiment than in the case of a lack of cooling (Study 1). On the chest (Figure 5a), the local skin temperatures were similar for both cooling variants, while on the back (Figure 5b), the local skin temperature was reduced by 4 °C in Study 3 (TE modules active) in comparison to Study 2 (just soaked heat sinks mounted) and by 7 °C in comparison to Study 1 (no cooling). The differences mentioned apply to the beginning of the initial phase of the tests; they decreased by between 2 °C to 3 °C at the end of the tests.

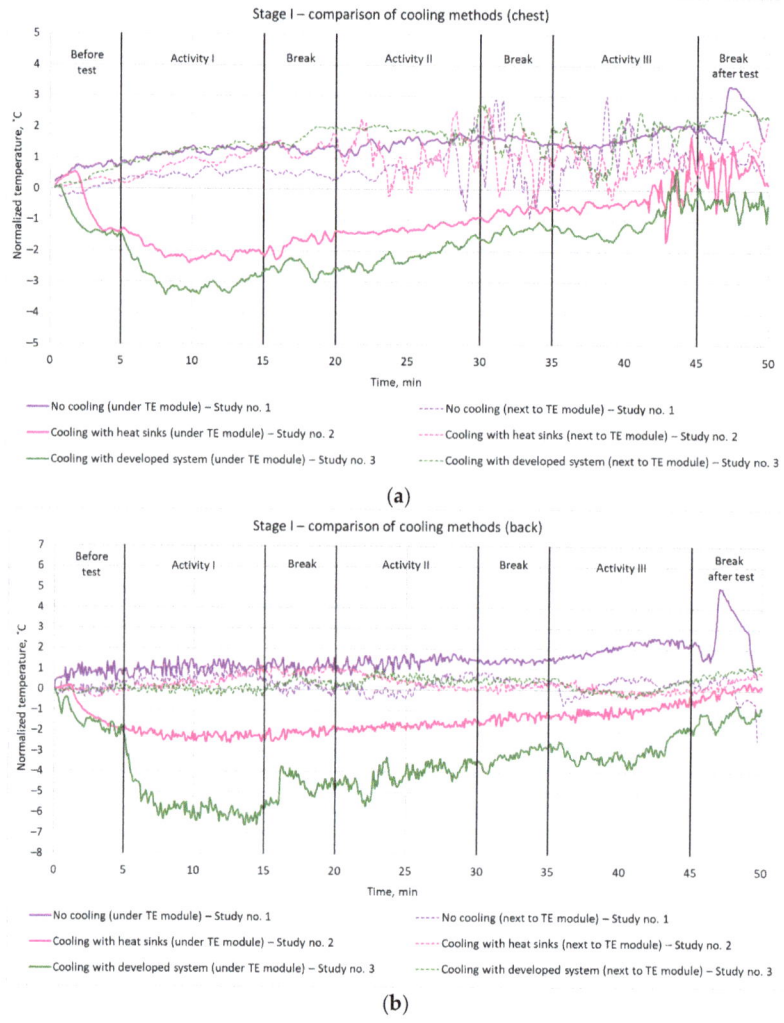

Figure 6. Normalized local skin temperature for different cooling variants (stage I, study numbering compliant with Table 1): (**a**) chest; (**b**) back.

On the chest (Figure 5a), a slightly lower temperature was observed with the TE modules off (Study 2) than when they were active (Study 3). However, this resulted just from a higher initial temperature in Study 3. The normalized temperature values shown in Figure 6a indicate that the cooling effect was in fact stronger with the TE modules on. In this latter case, the temperature drop was around 2 °C in the first 3 min of the test. On the back (Figure 6b), a greater drop in temperature was also observed with the TE modules active (Study 3). After 3 min from the start of cooling, a temperature drop of 4 °C was obtained under the TE module.

For all the tested variants, the measured temperatures increased throughout the entire period of the study. However, only with the TE modules active (Study 3), the temperature at the end of the test did not exceed the temperature measured at its beginning. This proves that active cooling using thermoelectric modules provides a more effective cooling than just using passive methods (soaked heat sinks).

3.2. Effect of Ambient Temperature (Stage II)

At the next stage, the PCS with soaked heat sinks mounted and TE modules active was tested at three different ambient temperatures: 25 °C, 30 °C and 35 °C (Table 1). The measured skin temperatures are shown in Figures 7 and 8.

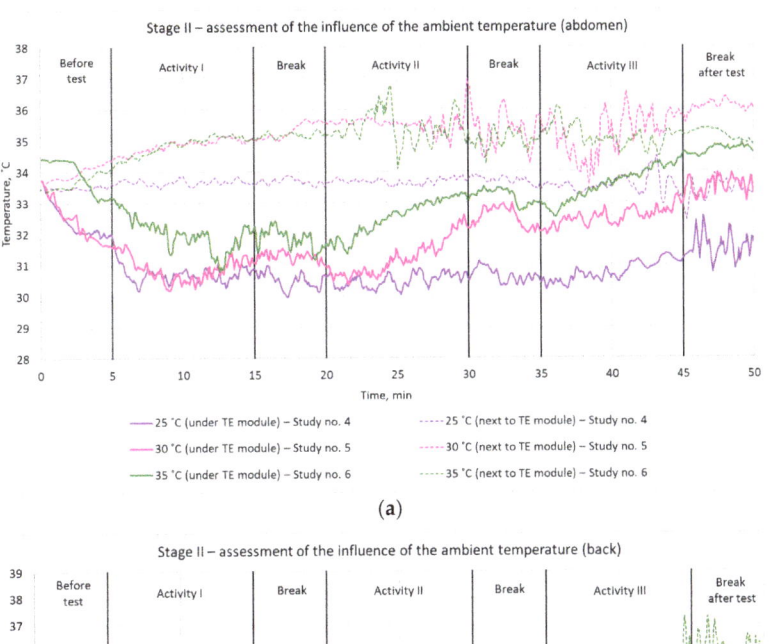

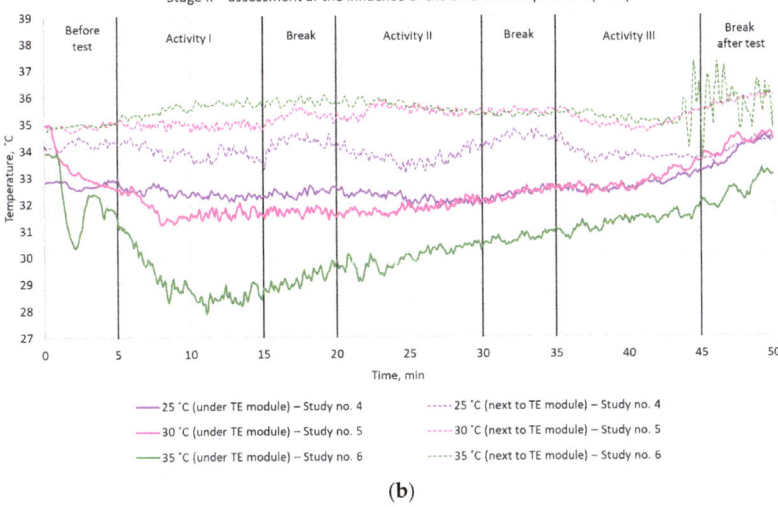

Figure 7. Local skin temperature under and next to a TE module for different ambient temperatures (stage II, study numbering compliant with Table 1): (**a**) abdomen; (**b**) back.

A cooling effect was observed with the use of the developed system for all three ambient temperatures. The lowest local skin temperature, of around 28 °C, was measured on the back during Activity I at an ambient temperature of 35 °C (Figure 7b). In this case, it was about 6 °C lower than the body temperature before the test, which is a very pronounced difference (Figure 8b).

On the abdomen, the cooling effect was similar to that on the back for all ambient temperatures in the first phase of the study (Figure 7a). In further phases, the largest local skin temperature drop with respect to its initial value, amounting to 3.5 °C, was obtained for an ambient temperature of 25 °C (Figure 8a).

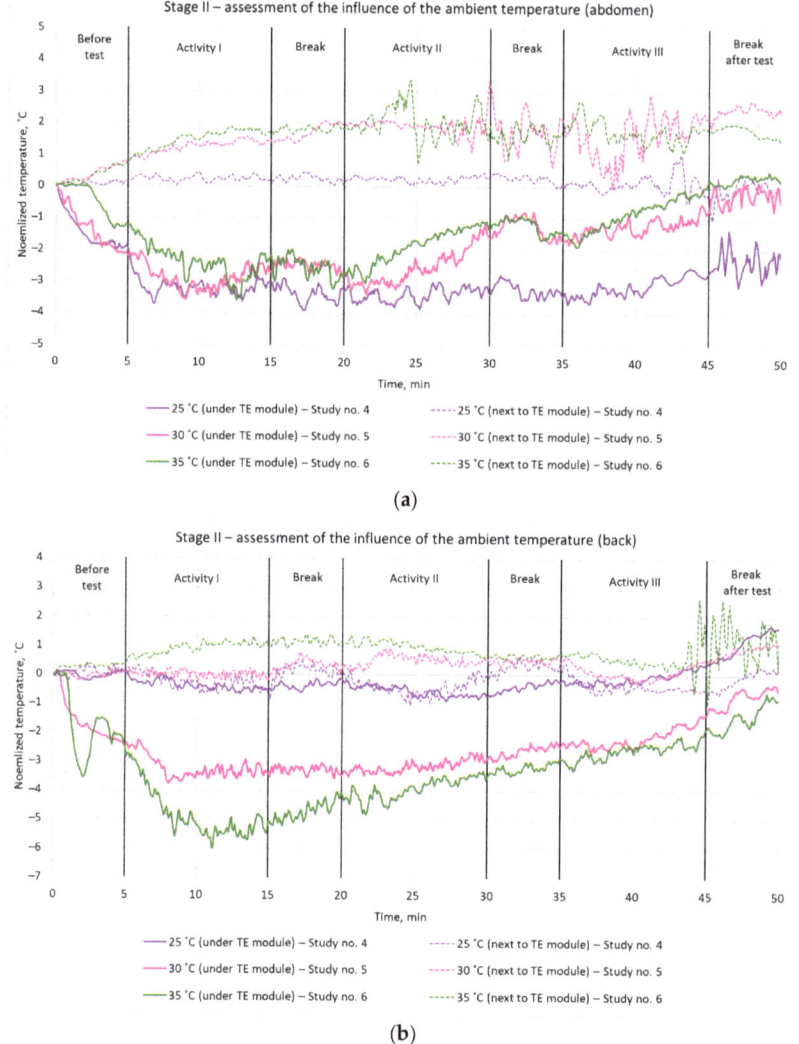

Figure 8. Normalized local skin temperature under and next to a TE module for different ambient temperatures (stage II, study numbering compliant with Table 1): (**a**) abdomen; (**b**) back.

In every case, the local skin temperature increased with time. This was due to the high intensity of physical exertion.

3.3. Power Consumption and Electronic Controller Efficiency

Table 3 contains the results related to electric power consumption and conversion in the system for Studies 4 through 7. They were therefore obtained for the same cooling variant (heat sinks mounted and TE modules active), while the ambient temperature varied. The durations of Studies 4 through 6 were almost equal (0.93 h to 0.94 h) while Study 7 was of a much longer duration (6.02 h). All the studies started with the battery fully charged.

Table 3. Electric power delivered to the TE modules and power drawn from the battery (study numbering as in Table 1).

Study	Ambient Temperature (°C)	Study Duration (h)	Average Battery Voltage (V)	Peak Battery Discharge Power (W)	Total Output Energy (Wh)	Total Energy Drawn from Battery (Wh)	Average Total Output Power (W)	Average Battery Discharge Power (W)	Average Controller Efficiency
4	25	0.93	20.54	14.9	5.68	6.36	6.09	6.83	0.89
5	30	0.93	20.55	14.9	7.46	8.17	8.05	8.82	0.91
6	35	0.94	20.50	15.1	8.42	9.13	8.93	9.67	0.92
7	30	6.02	19.66	14.9	52.15	56.84	8.67	9.45	0.92

It should be noted that the total output power recorded by the electronic controller is not exactly equal to the total electric supply power of the TE modules. This is due to the resistance of the connecting wires which was estimated to be approximately 0.27 Ω. This brings an additional power loss of 9.5% when two TE modules are supplied, whose equivalent resistance is 2.6 Ω. The possibility of reducing these losses is limited, due to the fact that any increase in the wire diameter decreases the comfort of the user.

On the other hand, power losses between the controller and the battery are negligible thanks to the much shorter distance, greater wire diameter and lower current (resulting from a higher converter input voltage as compared to its output one). Thus, the input voltage is, to a good approximation, equal to the battery voltage.

On the basis of the measurement results contained in Table 3, the battery power was calculated for each sample as

$$P_{bat}[n] = U_i[n] \times I_i[n]. \tag{9}$$

where U_i is the controller's input voltage, I_i is its input current and n is the sample number. Then, the energy drawn from the battery and the energy delivered to the converter's output were evaluated according to, respectively,

$$W_{bat} = \sum [P_{bat}[n] \times (t[n] - t[n-1])], \tag{10}$$

$$W_{o,tot} = \sum [P_{o,tot}[n] \times (t[n] - t[n-1])]. \tag{11}$$

where $P_{o,tot}$ is the recorded controller's total output power and t is the sample time stamp. Consequently, the respective average powers were

$$P_{bat(av)} = W_{bat}/\Delta t_{study}, \tag{12}$$

$$P_{o,tot(av)} = W_{o,tot}/\Delta t_{study}, \tag{13}$$

where Δt_{study} is the study duration.

Finally, the average power conversion efficiency of the electronic controller was evaluated as

$$\eta = P_{o,tot(av)}/P_{bat(av)}. \tag{14}$$

This overall average electronic controller efficiency includes both power losses in all the power converters and power consumption by the microcontroller, the display and auxiliary circuitry.

To investigate the effect of varying operating conditions of the controller, an instantaneous efficiency was also evaluated for the consecutive samples using the following formula:

$$\eta[n] = P_{o,tot}[n]/P_{bat}[n]. \tag{15}$$

In this case, only those time intervals were taken into account where neither the battery nor the output power varied, applying a limit of 5% variation sample-to-sample. This was because the reaction at the input of the controller (battery current) is delayed with respect to the output power, this resulting from the dynamic characteristics of switch-mode power converters [13]. Consequently, the result of Equation (15) is only physically meaningful in a steady state. The obtained overall instantaneous efficiency has been plotted in Figure 9 as a function of the total output power $P_{o,tot}$. It can be seen that the characteristic obtained does not depend on a particular study, thus on ambient temperature or test duration.

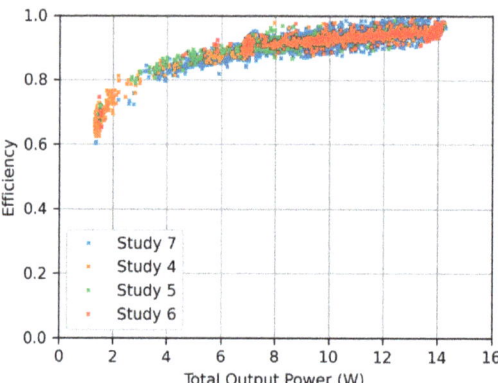

Figure 9. Overall instantaneous efficiency of the electronic controller as a function of its total output power (study numbering as in Table 1).

4. Discussion

4.1. Cooling Efficiency

The results of the study confirm that the developed PCS based on TE modules is able to provide efficient cooling with a temperature drop of about 3.5 °C. This result is coherent with the results obtained by Hong et al. [9] who also worked on the implementation of flexible TE modules in PCS; however, their solution was not intended for physical workers.

According to our observations, cooling efficiency of the PCS with TE modules and evaporative heat sinks is highly dependent on both the heat capacity and the rate of water evaporation from the heat sink, as well as on the fitting of the TE module to the user's body. The large temperature drop prior to the beginning of the test results from the temperature of the water absorbed by the heat sink being lower than the hot side of the TE module. When the temperature of the water and that of the TE module's hot side become equal, the cooling efficiency is decreased. Nevertheless, even then the evaporation of water from the soaked heat sink still provides a cooling effect whose intensity is stable in time. Following Cotter et al. [14], local cooling of about (−4) °C can be considered as mild, and the results obtained with our TE-based PCS were quite similar to the effect provided by the personal liquid cooling vest described by Xu et al. [15] and are within the range of reduction of the local skin temperature (from 1.1 to 7.4 °C) achieved by Song and Wang [16] by means of PCS with PCMs and small air ventilation fans.

The obtained results also indicate the importance of a proper fitting of the TE modules to the user's body. While all modules were supplied with nearly the same electric power, differences are observed between the modules in the resulting local skin temperature. The particular design of the clothing enables better fitting of those TE modules that are located on the back. That is why a stronger cooling effect is observed there than on the chest.

Without any cooling (no heat sinks and TE modules turned off), temperatures under TE modules are slightly higher than those next to them throughout the entire test duration, both on the chest and on the back. This is caused by the presence of an additional layer—a TE module—at the place of measurement. As the blouse's material was highly air-permeable, this resulted in lower temperatures outside the TE modules. The research outcomes in the above-discussed field related to the influence of water evaporation from the flexible heat sinks in the TE-based PCS as well as the fitting of the TE modules to the human body on the PCS cooling efficiency have not been considered up to now. Therefore, they indicate a completely new research direction in environmental engineering that will be explored further.

4.2. Energy Consumption and Operating Time

As it can be seen in Table 3, the peak power drawn from the battery is about 15 W for any study. This value results from the power limitation of 2 W per TE module that is activated each time the temperature controller requests a maximum module supply power.

On the other hand, both the average battery power and the average output power rise with the ambient temperature increasing. This is expected as more heat has to be evacuated from the body in order to maintain a comfortable body temperature, which in turn requires a higher temperature difference to be produced between the ambient temperature and the body. The average battery power has increased by 42% with the ambient temperature rising from 25 °C to 35 °C. This is a very pronounced change, affecting the required capacity of the battery and, consequently, the size, weight and cost of the latter.

The average output power is higher in Study 7 than in Study 5 despite the ambient temperature being the same. This may result from the efficacy of heat sinks decreasing over the long-time scale due to evaporation of water therefrom. To compensate for this, the electric supply power of the TE modules must be increased.

Based on the average battery power of 9.45 W obtained for the long-time study, a battery with a capacity of 10.2 Ah and a nominal voltage of 3.7 V (lithium-ion technology) is sufficient to supply the system for 4 h, which corresponds to a standard power bank whose weight is typically about 230 g. A battery of 20.4 Ah would be required to extend the operating time to 8 h.

In papers where similar wearable cooling systems have been described, power consumption is rarely analyzed. In [17], the ambient temperature varied between 27 °C and 35 °C, while the power consumption was between 10 W and 15 W, which is higher than for the system presented here (supposing that the value of 10 W corresponded to 27 °C, the difference may be estimated by linear interpolation as 31%). This may have resulted from the use of a fan and a water pump which additionally contributed to the overall weight of 680 g for an operating time of only 2 h. The total weight of the system presented in this paper is similar at 650 g with a battery of 3.4 Ah and 18.2 V (nominal values), allowing it to operate for 6.4 h (3.2 times longer) under the worst-case temperature of 35 °C.

The body temperature regulator described in [18] consumed 5.58 W (0.05 W in standby) with just a single TE module versus seven modules in this work. However, it seems that that system was not tested on humans, so the practical effect of this supply power on body temperature is unknown.

The electric energy conversion efficiency of the developed dedicated electronic controller was high. According to Table 3, its average value over an entire study, as defined by Equation (14), was 0.89 to 0.92. This efficiency decreased with decreasing output power. This effect can be seen more clearly in Figure 9, where the instantaneous efficiency has been plotted, as defined by Equation (15). It may be explained by an important contribution of those components of the power consumption that are independent of or weakly dependent on the output power, such as the supply power of the control circuitry, especially the microcontroller, the display and the SD card.

For high output power, those components become dominant that depend on the output current, including losses in power converters' components, e.g., transistors and inductors. As these losses have been minimized by careful component selection, the efficiency then reaches 0.95, which is identical with the result given in [17]. The nature of the relationship observed is favorable for a battery-supplied system, as it keeps the absolute power loss low: the efficiency is high when the output power is high, and it is only low when the output power is low as well.

It should be noted that much of the control circuitry power consumption is due to those functions that were necessary for research purposes, such as the OLED display, the

SD card and the battery monitor. These features may be eliminated in a final version of the system, thus increasing the overall efficiency and reducing the energy drawn from the battery.

5. Conclusions

The laboratory tests of the developed personal cooling system carried out with human participation allowed the efficiency of active cooling using TE modules to be compared to passive cooling using only soaked heat sinks. They also made it possible to investigate the effect of the ambient temperature. The use of TE modules provided an efficient reduction of the skin temperature during an entire experiment of about 1 to 6 h involving moderate physical intensity.

It follows from the results obtained that a proper fitting of the clothing to the user's silhouette is required to efficiently draw the excess heat out of the human body. With those TE modules that better adhere to the body, a much higher temperature difference can be achieved. Moreover, in the first stage of the research, we proved a favorable effect of the use of evaporative heat sinks (Study 2) as well as a synergic effect of the use of TE modules with those heat sinks (Study 3). Furthermore, without effectively drawing heat from the hot side of the TE modules, they will not provide an expected cooling function. The tests conducted in different ambient temperatures show similar differences in local skin temperature between cooled and uncooled locations. This suggests that ambient temperature has no significant effect on the cooling efficiency of the proposed PCS, at least in the range between 25 °C and 35 °C.

The experimental results demonstrate that power losses in the power electronic converters are relatively low, enabling an average efficiency of 0.90 and a maximum one of 0.95 (for a maximum TE module electric power), which is the same as in a similar system. On the other hand, the overall electronic controller efficiency includes the power consumption by the control circuitry that becomes more prominent at low output power values, leading to a decrease in the efficiency. However, much of this consumption is related to functions that were only implemented for research purposes and may be eliminated in a future version of the system to decrease the power demand. This will allow the battery size and weight to be reduced or the operating time to be increased.

The research stage described in the present paper was focused on the cooling efficiency and energy efficiency aspects of system operation. The conclusions from this research will be used in the development of an improved PCS prototype. The main limitations of the performed tests are related to the involvement of only one participant. Nevertheless, this has been sufficient to reliably assess the PCS in respect of its cooling performance and power consumption thanks to a repeatable testing methodology adopted. To formulate wider conclusions related to the evaluation of PCS efficiency, it is planned to test the prototype with multiple end users. Those further tests will involve workers carrying out physical operations characteristic for their real workplaces. This will enable us to study the ease of use, ergonomics as well as functionality of the developed PCS and its compatibility with other equipment used during routine work. Moreover, based on further experiments, a decision on a potential limitation of the number of TE modules will be made to make PCS more cost-effective and increase its potential for successful placement on the market.

Author Contributions: Conceptualization, A.D.; data curation, A.D., M.K., Ł.S. and B.P.; investigation, A.D., M.K., Ł.S. and B.P.; methodology, A.D.; resources, A.D.; supervision, A.D.; validation, A.D., M.K., Ł.S. and B.P.; visualization, M.K., Ł.S. and B.P.; writing—original draft, A.D., M.K., Ł.S. and B.P.; writing—review and editing, A.D., M.K., Ł.S. and B.P. All authors have read and agreed to the published version of the manuscript.

Funding: This paper is published and based on the results of a research task carried out within the scope of the fifth stage of the National Programme "Improvement of safety and working conditions" supported from the resources of the National Centre for Research and Development (task no. III.PB.09), entitled "Development of the protective clothing with an active cooling function based on the thermoelectric effect (Peltier modules)". The Central Institute for Labour Protection—National Research Institute is the Programme's main coordinator.

Institutional Review Board Statement: Not applicable.

Informed Consent Statement: An informed consent statement was signed by the participant involved in the laboratory tests.

Data Availability Statement: The data presented in this study are available on request from the corresponding author.

Conflicts of Interest: The authors declare no conflict of interest.

References

1. Council of the European Union. *Council Directive 89/391/EEC of 12 June 1989 on the Introduction of Measures to Encourage Improvements in the Safety and Health of Workers at Work*; Council of the European Union: Brussels, Belgium, 1989.
2. Sajjad, U.; Hamid, K.; Rehman, T.U.; Sultan, M.; Abbas, N.; Ali, H.M.; Imran, M.; Muneeshwaran, M.; Chang, J.-Y.; Wang, C.-C. Personal thermal management—A review on strategies, progress, and prospects. *Int. Commun. Heat Mass Transf.* **2022**, *130*, 105739. [CrossRef]
3. Al Sayed, C.; Vinches, L.; Dupuy, O.; Douzi, W.; Dugue, B.; Hallé, S. Air/CO_2 cooling garment: Description and benefits of use for subjects exposed to a hot and humid climate during physical activities. *Int. J. Min. Sci. Technol.* **2019**, *29*, 899–903. [CrossRef]
4. Bartkowiak, G.; Dabrowska, A.; Marszalek, A. Assessment of an active liquid cooling garment intended for use in a hot environment. *Appl. Ergon.* **2017**, *58*, 182–189. [CrossRef] [PubMed]
5. Lu, Y.; Wei, F.; Lai, D.; Shi, W.; Wang, F.; Gao, C.; Song, G. A novel personal cooling system (PCS) incorporated with phase change materials (PCMs) and ventilation fans: An investigation on its cooling efficiency. *Therm. Biol.* **2015**, *52*, 137–146. [CrossRef] [PubMed]
6. Qiao, Y.; Cao, T.; Muehlbauer, J.; Hwang, Y.; Radermacher, R. Experimental study of a personal cooling system integrated with phase change material. *Appl. Therm. Eng.* **2020**, *170*, 115026. [CrossRef]
7. Ding, J.; Zhao, W.; Jin, W.; Di, C.; Zhu, D. Advanced Thermoelectric Materials for Flexible Cooling Application. *Adv. Funct. Mater.* **2021**, *31*, 2010695. [CrossRef]
8. Wei, H.; Zhang, J.; Han, Y.; Xu, D. Soft-covered wearable thermoelectric device for body heat harvesting and on-skin cooling. *Appl. Energy* **2022**, *326*, 119941. [CrossRef]
9. Hong, S.; Gu, Y.; Seo, J.K.; Wang, J.; Liu, P.; Meng, Y.S.; Xu, S.; Chen, R. Wearable thermoelectrics for personalized thermoregulation. *Sci. Adv.* **2019**, *5*, eaaw0536. [CrossRef] [PubMed]
10. Li, Z.; Zhang, M.; Yuan, T.; Wang, Q.; Hu, P.; Xu, Y. New wearable thermoelectric cooling garment for relieving the thermal stress of body in high temperature environments. *Energy Build.* **2023**, *278*, 112600. [CrossRef]
11. Dąbrowska, A.; Kobus, M.; Starzak, Ł.; Pękosławski, B. Analysis of Efficiency of Thermoelectric Personal Cooling System Based on Utility Tests. *Materials* **2022**, *15*, 1115. [CrossRef] [PubMed]
12. Firdaus, A.Z.A.; Zolkifly, M.Z.; Syahirah, K.N.; Normahira, M.; Aqmariah, S.N.; Ismail, I.I. Design and development of controller for Thermoelectric cooler system. In Proceedings of the 2015 IEEE International Conference on Control System, Computing and Engineering (ICCSCE), Penang, Malaysia, 27–29 November 2015; pp. 264–268. [CrossRef]
13. Maniktala, S. *Switching Power Supplies A–Z*, 2nd ed.; Newnes: Oxford, UK, 2012. [CrossRef]
14. Cotter, J.D.; Taylor, N.A.S. The distribution of cutaneous sudomotor and alliesthesial thermosensitivity in mildly heat-stressed humans: An open-loop approach. *J. Physiol.* **2005**, *565 Pt 1*, 335–345. [CrossRef] [PubMed]
15. Xu, J.; Chen, G.; Wang, X.; Chen, Z.; Wang, J.; Lu, Y. Novel Design of a Personal Liquid Cooling Vest for Improving the Thermal Comfort of Pilots Working in Hot Environments. *Indoor Air* **2023**, *2023*, e6666182. [CrossRef]
16. Song, W.; Wang, F. The hybrid personal cooling system (PCS) could effectively reduce the heat strain while exercising in a hot and moderate humid environment. *Ergonomics* **2016**, *59*, 1009–1018. [CrossRef] [PubMed]
17. Itao, K.; Hosaka, H.; Kittaka, K.; Takahashi, M.; Lopez, G. Wearable Equipment Development for Individually Adaptive Temperature-conditioning. *J. Jpn. Soc. Precis. Eng.* **2016**, *82*, 919–924. [CrossRef]
18. Chen, Y.-S.; Chuang, Y.-R.; Chiang, H.-H.; Chen, Y.-L. The Design of an Automatic Body Temperature Regulator. In Proceedings of the 2019 IEEE International Conference on Consumer Electronics—Taiwan (ICCE-TW), Yilan, Taiwan, 20–22 May 2019; pp. 1–2. [CrossRef]

Disclaimer/Publisher's Note: The statements, opinions and data contained in all publications are solely those of the individual author(s) and contributor(s) and not of MDPI and/or the editor(s). MDPI and/or the editor(s) disclaim responsibility for any injury to people or property resulting from any ideas, methods, instructions or products referred to in the content.

Article

Employing the Peltier Effect to Control Motor Operating Temperatures

Stephen Lucas [1,2,*], Romeo Marian [1], Michael Lucas [1], Titilayo Ogunwa [1] and Javaan Chahl [1,3]

1. UniSA STEM, University of South Australia, Mawson Lakes, SA 5095, Australia
2. Lucas Solutions Pty. Ltd., Hazelwood North, VIC 3840, Australia
3. Joint and Operations Analysis Division, Defence Science and Technology Group, Melbourne, VIC 3207, Australia
* Correspondence: stephen.lucas@mymail.unisa.edu.au

Citation: Lucas, S.; Marian, R.; Lucas, M.; Ogunwa, T.; Chahl, J. Employing the Peltier Effect to Control Motor Operating Temperatures. *Energies* 2023, *16*, 2498. https://doi.org/10.3390/en16052498

Academic Editor: Diana Enescu

Received: 24 January 2023
Revised: 27 February 2023
Accepted: 4 March 2023
Published: 6 March 2023

Copyright: © 2023 by the authors. Licensee MDPI, Basel, Switzerland. This article is an open access article distributed under the terms and conditions of the Creative Commons Attribution (CC BY) license (https://creativecommons.org/licenses/by/4.0/).

Abstract: Electrical insulation failure is the most common failure mechanism in electrical machines (motors and generators). High temperatures and/or temperature gradients (HTTG) are the main drivers of insulation failure in electrical machines. HTTG combine with and augment other destructive effects from over-voltage, to voltage transients, overload and load variations, poor construction techniques, and thermal cycling. These operating conditions cause insulation damage that leads to electrical insulation failure. The insulation failure process is greatly accelerated by pollutants and moisture absorption. A simple and robust way to reduce HTTG and moisture adsorption is by maintaining constant internal temperatures. The current method to maintain elevated internal temperatures and reduce condensation issues is by internal electrical heating elements. This paper examines the effectiveness of applying thermoelectric coolers (TECs), solid-state heat pumps (Peltier devices), as heaters to raise a motor's internal temperature by pumping heat into the motor core rather than heating the internal air. TEC technology is relatively new, and the application of TECs to heat a motor's internal volume has not previously been explored. In this paper, we explore the hypothesis that TECs can pump heat into a motor when out of service, reducing the HTTG by maintaining high winding slot temperatures and eliminating condensation issues. This paper describes a test motor setup with simple resistive heating (traditional method), compared with the application of TECs with heat sinks, heat pipes, and a water circulation heat exchanger, to gauge the capability of TECs to heat the inner core or winding area. In this paper, we demonstrate the full integration of TECs into a motor. The results show that each of the systems incorporating the TECs would effectively pump heat into the core and keep the winding hot, eliminating condensation issues and water ingress due to thermal cycling.

Keywords: electric motor; condensation; insulation; thermoelectric device; thermal management

1. Introduction

Electrical motors are now incorporated into every facet of today's life and primarily convert electrical energy to mechanical energy. Their makeup consists of steel cores with insulated electrical windings, bearings, rotors, and couplings. A key material affecting the life of a motor is the electrical insulation. The electrical insulating material applied to the winding conductors used in electrical motors varies from simple enamel coatings to layers of glass/mica tape impregnated with resins. It can also consist of layers of specialized insulating material such as Nomex or Kapton. The type of insulation and the number of layers applied depend on the operating voltages and thermal conditions. The more exotic the material, the higher the costs. Its purpose is to insulate the turn conductors from each other and from the core. All insulation systems are susceptible to mechanical and thermal damage [1], and their ability to maintain their insulating property can be greatly affected by pollutants such as dirt, oil, and moisture [2].

The insulation system for small motors up to 300 kW is primarily made from enamel resins, and for large motors, a resin-impregnated fibreglass carrier with mica is normally used. When motors are in service, this insulation is affected by several operating conditions. High temperatures (HT) and temperature gradients (TG) are key issues affecting insulation integrity, but normal operating conditions such as thermal cycling, load variations, voltage fluctuations, system harmonics, and daily environmental temperature changes cause insulation aging. For motors greater than 50 kW, the air gap flow heating phenomena [3] also affect insulation life. Normal maximum running temperatures by design are set by the class of insulation used. If temperatures exceed the class temperature by 10 °C, the insulation life will be halved. Variations in the operation temperature result in varying dimensional changes to the windings, insulation, core, and wedging systems. The mechanical stresses on the insulation due to different thermal expansion rates damage the insulation and result in surface damage and microcracking of the insulation [4,5]. This then enables moisture absorption and higher leakage currents [6].

New motors with new windings and insulation are nearly waterproof, as the insulation systems have little surface damage and absorb little moisture. Motors that have been in service for a long time are extremely hygroscopic [7,8], and a day of high humidity can result in an extremely low insulation resistance (IR) value [9]. Electrically testing a winding's IR will show the serviceability of the insulation. Standards give a minimum IR value of $2E + 1$ MΩ, where E is the line voltage in kV [10].

When motors or generators are in service, they are hot, and moisture issues are rare [11]. Motors that are continually running for long periods of time suffer less from mechanical stresses, as the operational temperatures are more stable, compared to motors that are frequently started and stopped. Base-load hydro generators are good examples of electrical winding with long lives. Pumps or pressurising systems are good examples of motors that have short lives.

In all cases, as the insulation ages, electrical insulation systems become more susceptible to moisture ingress and failure [8,12]. Even motors stored for future use suffer from moisture ingress due to normal daily thermal cycling [13]. When internal air heats, it expands, causing it to bleed out of the motor. When it cools and contracts, it pulls air in with moisture from outside. Continued thermal cycling concentrates the internal moisture levels. This process affects in-service motors as well as motors stored as spares for operational backup. Over time, this daily thermal cycling can result in many litres of water being trapped inside the motor.

Because of the design of a motor, there is a large amount of iron in the stator and rotor cores, and they also have substantial frames. Consequently, a large mass results in a time delay in temperature stabilization with the external environment. The resulting temperature difference can result in internal condensation with the internal surfaces becoming wet and the electrical insulation absorbing high levels of moisture [8]. This high level of moisture contamination can lead to catastrophic insulation failure [14]. With hot days, cool nights, and the thermal inertia inside the machine, condensation is an issue. The point where moisture in the internal air will condense onto surfaces [15,16] or above daily temperatures [17] is called the dew point. As shown in Figure 1 [2], with an air temperature of 27 °C and a humidity of 60%, the dew point is 18 °C, and at 80% humidity, the dew point is 23 °C. In normal conditions with cold nights and the air temperatures quickly rising when the sun rises, the conditions for internal condensation are easily reached, and internal heaters often fail to prevent internal condensation.

The degree of protection required and the practical systems to maintain motors above the dew point are dependent on a number of factors. Of immediate consideration is whether the motor is cycling in and out of service or whether it is being stored from the short- to the long term. If it is being stored, the environmental conditions become very important with internal temperature-controlled warehouses being far more protective than outside storage. To help eliminate internal condensation and moisture absorption, the internals of a motor need to be kept above the dew point [7]. The most common method employed to achieve this is the installation of resistive heaters mounted internally. Another system of heating heats the winding by circulating a low current via electronic supplies [7].

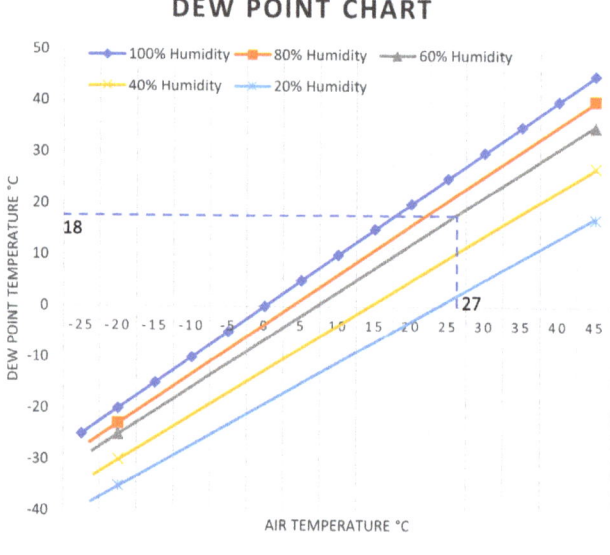

Figure 1. The relationship between the air temperature, the dew point, and the humidity (adapted from [2]).

For storage, portable electric heaters inside a storage container, or for large machines, temporarily installed space heaters can be used [13]. Partial disassembly of the motor eliminates an internal environment that can result in condensation [17]. The use of waterproof bagging with several bags of desiccant (silica gel) can also be used [17]. Another option is called cocooning, sometimes used by the military, where the motor is protected and then covered with a layer of plastic sheeting; then, a continuous layer of plastic film is sprayed directly onto the motor surface [17]. Moreover, the process of slowly raising the motor temperatures while circulating dry air can remove upper level moisture contamination [18] before placing the motor back into service.

The heating methods currently employed, as described above, have difficulty in keeping motors above the dew point [2]. The resistive heaters commonly used are limited because they heat the internal air and have difficulty coping with rapid external temperature changes. Winding heating systems require extra equipment that adds to the installation and maintenance costs and is rarely applied. The cocooning option used by the military is for extreme conditions, and when in place, it is regularly renewed, increasing maintenance costs. Storage systems try to limit thermal cycling, but this is difficult to achieve.

This research investigates the application of thermoelectric coolers (TECs) to heat a motor to decrease the thermal gradients and stresses, eliminate the internal condensation issues, and stop the water absorption. The TEC module is a solid-state heat pump, and the direction the heat is pumped depends on the supply polarity applied to the module. The aim of this study is to set the TEC up to pump heat into the core to determine whether the inner core temperatures at the winding position can be controlled, providing the opportunity to keep the winding and internals hot. The TECs were incorporated into three basic systems to measure their impact on the heating system. This would help to control the condensation issues and reduce the thermal cycling stresses on the insulation and bracing systems. It is not a comparison between systems, as the system configuration requirements are dependent on the load and the environmental conditions and are different for each motor system. Any system that can improve the life of a motor and reduce the life cycle costs is worth exploring. The introduction of new technology often leads to additional environmental problems [19], but the introduction of the TEC and the ability to directly pump heat into the core will reduce the temperature differences, reducing the mechanical stresses and moisture issues without impacting other systems.

This research used a standard electrical motor integrated with sensors to monitor the temperatures at various locations using TECs to control the internal temperatures. The use of TECs to control internal temperatures has not been previously tried, and this was the first attempt to incorporate these low-cost devices into a fully instrumented motor working in standard operating conditions to improve the reliability and reduce the life cycle cost.

Section 2 explains the TEC devices available and their performance. Section 3 describes the test motor and each of the four heating setups (resistors, heat sinks, heat pipes, and a water heat exchanger) with the results from the four heating systems. Section 4 discusses the obtained results, and Section 5 concludes with the findings of the study.

2. Thermoelectric Cooling Device (Peltier Module)

The makeup of a thermoelectric cooling device is a combination of N and P type material between two ceramic plates. The most common and economical material is bismuth telluride (Bi_2Te_3) or lead telluride (PbTe). These are low temperature materials that limit the operating range of the TEC. However, the base materials can be alloyed with other materials to improve the performance at various temperatures [20]. Thus, by alloying the base materials with other materials, the temperature ranges can be significantly increased to match the application.

The expected life of these units is 200,000 h, but care must be taken to not exceed the maximum ratings. There are also a variety of these units; for example, the TEC1-12715 (A$6.08) is rated for a maximum temperature of 70 °C, and the TEC1-12715HTS (A$33.58) has a rating of 200 °C. The cost increases in proportion to the increased temperature range.

When a current passes through the TEC, heat is transferred from one plate to the other, as shown in Figure 2, and energy is pumped from one plate to the opposite plate. Today, these modules can be found in portable water coolers or heated seats in the car industry. Figure 2 shows a schematic diagram [21], and Figure 3 provides a view of the internal constructional layout.

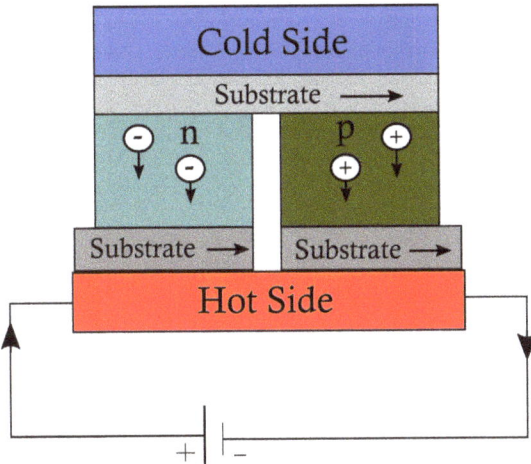

Figure 2. Schematic diagram of a typical thermoelectric module (adapted from [21]).

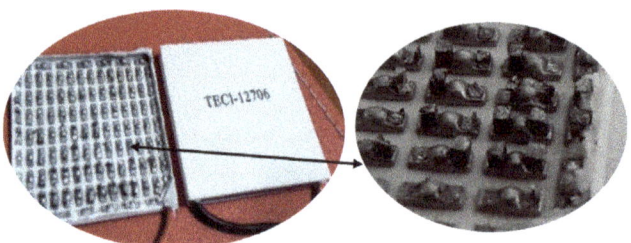

Figure 3. The Peltier (cooling module) arrangement.

The TEC data sheets are based on the cold plate being held at 0 °C. In application, the operation of the cold plate is likely to be subjected to climatic and industrial variations in temperature, so, six easily obtained TECs were tested to develop a better understanding of their operational performance. The test setup in Figure 4, also shown in Figure 9 in [22], was developed to allow the TECs to be tested and the performance characteristics to be compared.

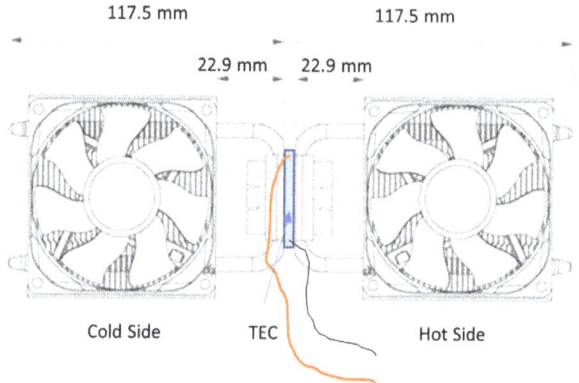

Figure 4. The TEC test layout.

Each of the TECs was installed into the test rig in turn, and the current was increased in 0.5 A steps until its rating was reached. A plot of the TEC current rating against the temperature is shown in Figure 5. As can be seen, the temperature gains were linear in proportion to the maximum TEC current rating.

The thermoelectric cooler (Peltier Module) chosen for this project was the TEC1-12706 (Figure 3). This unit was the lowest cost unit (AUD 3.21 per module) readily available online with operational characteristics suitable for experimentation. It had a 60 W cooling capability at nominally 12 V and 6 A. Further information and specification data are readily available in [23,24].

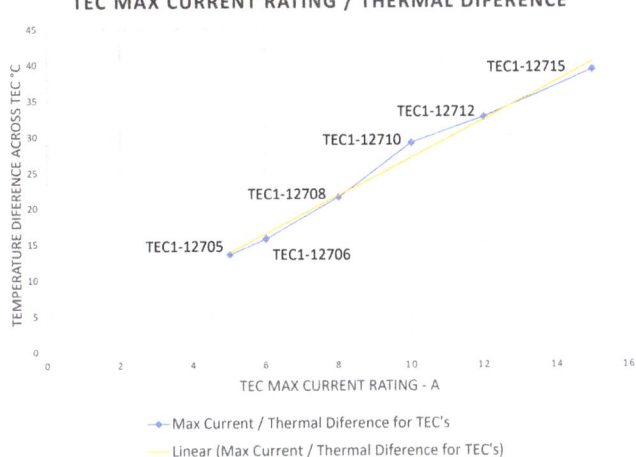

Figure 5. The thermoelectric cooler (TEC) max current versus the temperature of the tested TECs.

3. Experimental Test on a Motor and the Results

Previous research [22] by the authors used TECs to heat a core sample from a section of a motor's inner core encompassing the winding slot area. This paper explores the previously untried application of TECs to heat a single phase 3/4HP (559 W) induction test motor. Four systems were employed namely: resistors, heat sinks, heat pipes, and a water heat sink. This paper is not a comparison between the systems but rather an application of each system to gauge its effectiveness to heat up the motor from a stabilized room temperature.

A motor with the same arrangement as a standard induction motor was selected as the test motor. The motor was dismantled and serviced (cleaned and varnished for mechanical stability, with new bearings installed), and eight new thermocouples were added to the windings as in Figure 6, to monitor and record the winding temperatures when under test. The slot area is a critical thermal area, and all data presented in this paper were taken from slot 13 (CH02, see Figure 6), as a thermocouple was inserted into the centre of the slot.

The test motor was switched off and allowed to stabilize at room temperature. There were four different configuration arrangements:

1. Resistive heating;
2. TECs with small heat sinks;
3. TECs with heat pipe coolers;
4. TECs with water circulation heat sinks.

With the motor at room temperature, the heating power supply was applied to each system. The polarity to the TECs was established so that the TEC pumped heat into the motor (the side against the motor became hot). The current supply to the configuration was adjusted in 0.5 A steps, with the input power to the resistor heaters and the TECs, and the resulting temperatures were recorded for each of the applied systems.

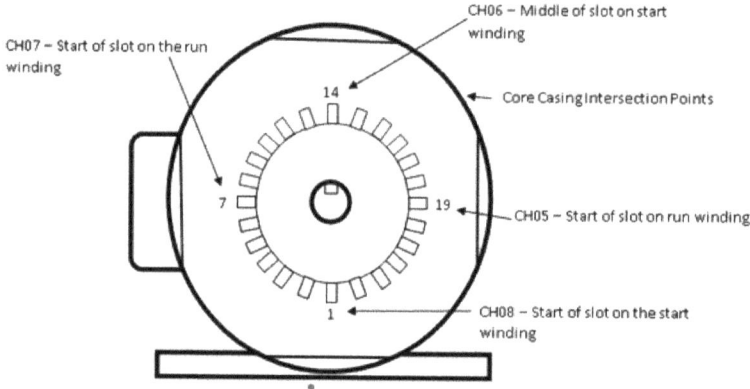
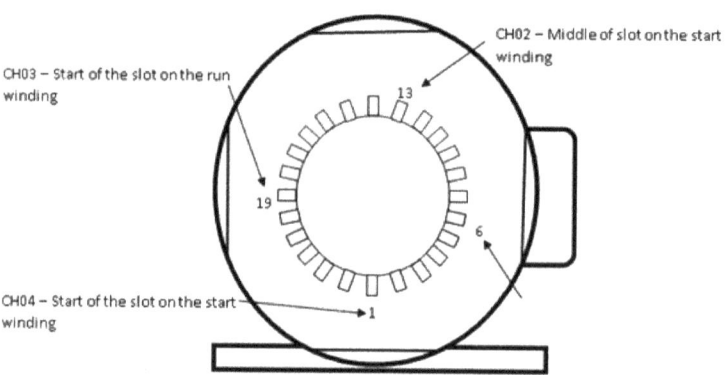

Figure 6. Schematic showing the placement of the thermocouples on the motor being tested. (**Top**): Front view. (**Bottom**): Back view.

3.1. Resistive Heating

The resistive heater system is the most common heating method in motors today. The resistive heating system used in this paper comprised three resistor heating banks attached to the outer surface of the motor (see Figure 7). This arrangement heated the motor, the core, and the slot where the stator windings were inserted. Although applied in a different manner, as they are normally installed inside the motor under the windings, they did raise the winding temperatures. The slot, casing, and resistance bank temperatures at various total power inputs to the resistors were recorded and plotted in Figure 8. At a nominal 50 W input, the slot temperatures reached 32.1 °C as shown in Figure 8, attaining thermal stability after 5 h.

Figure 7. Resistive heating of the test motor.

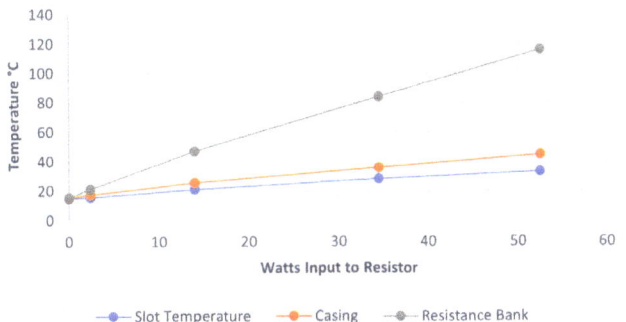

Figure 8. The graphed results of the heating with the resistors.

3.2. TECs with Small Heat Sinks

With this configuration, 14 TECs with small heat sinks were attached to the outer surface of the motor as shown in Figure 9. The TECs with small heat sinks had a small footprint, which allowed the largest application of the units to the test motor. This enabled a large coverage (14 TECs) of the casing, which minimized the losses from exposed surfaces. The mounting arrangement of the TECs was via a curved aluminium adapter plate (Figure 10), and the curve matched the diameter of the motor casing. Figure 11 shows a plot of the slot, casing, and heat sink temperatures at various total power inputs to the TECs. For a power less than 60 W, the temperature of the heat sink dropped below the atmospheric temperature due to the extraction of the heat from the environment and pumping it into the motor, coupled with the small area of the heat sinks. The temperature of the heat sink increased slightly when the input power was above 60 W due to the Joule effect. At the nominal 50 W input, the winding slot in the core reached 34.95 °C, attaining thermal stability after 4.5 h (Figure 11).

Figure 9. The setup of the heating of the test motor with the TECs and small heat sinks.

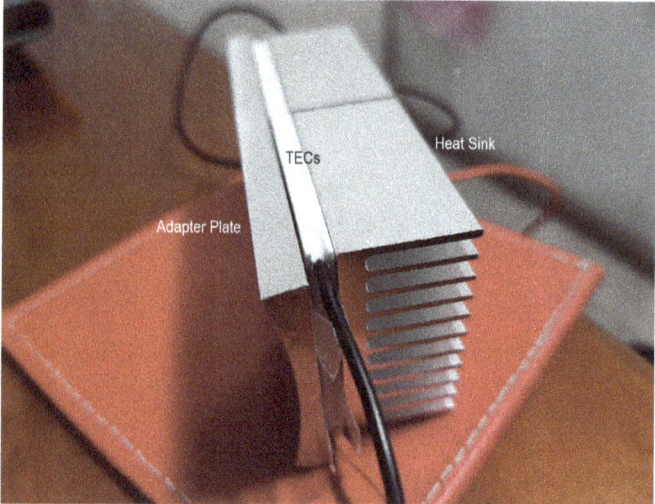

Figure 10. An adapter plate with the TEC and heat sink setup.

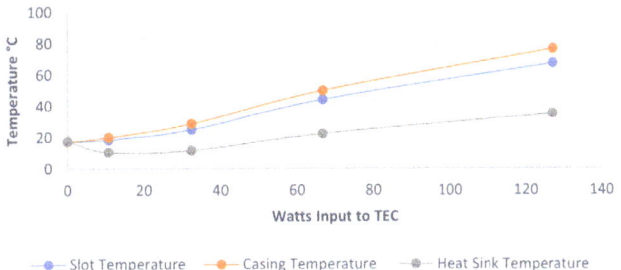

Figure 11. The graphed results of the heating with the TECs and small heat sinks.

3.3. TECs with Heat Pipe Coolers

This configuration consisted of TECs with heat pipe coolers attached to the outer surface of the motor (see Figure 12). The system utilizing the heat pipe heat exchangers [25,26] took up room; hence, only three TECs were used in this configuration. The recorded slot, casing, TEC top side, and heat sink temperatures at various total power inputs to the TECs are shown in Figure 13. At the nominal 50 W input power to the inner core area, the motor slots reached 42.93 °C, attaining thermal stability after 5 h (Figure 13).

The system, because of the greater area of the heat sinks, was more efficient.

Figure 12. The setup of the heating of the test motor with the TECs and heat pipe coolers.

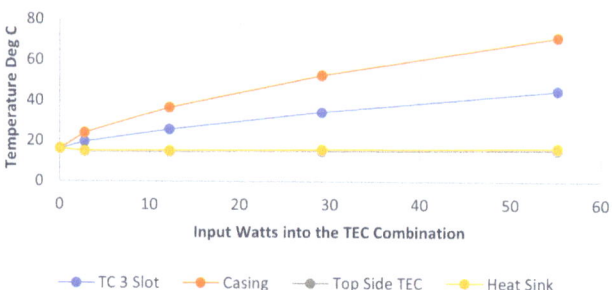

Figure 13. The graphed results of the heating with the TECs and heat pipe coolers.

3.4. TECs with Water Circulation Heat Sinks

The configuration considered here comprised TECs with water circulation heat sinks attached to the outer surface of the motor (see Figure 14). The heating system consisted of 10 TECs with water heat sinks and water running at 12 °C. This system worked similarly to the TECs with the small heat sinks. The recorded slot, casing, heat exchanger, and water outlet temperatures at various total power input to the TECs are plotted in Figure 15. With the water temperature at a low 12 °C at the heat source, the TEC combination performance was slightly limited, but at the 50 W nominal input setting, it raised the inner core windings to 35.48 °C, attaining thermal stability after 5 h (see Figure 15).

Figure 14. The setup of the heating of the test motor with the TECs and water cooled heat sinks.

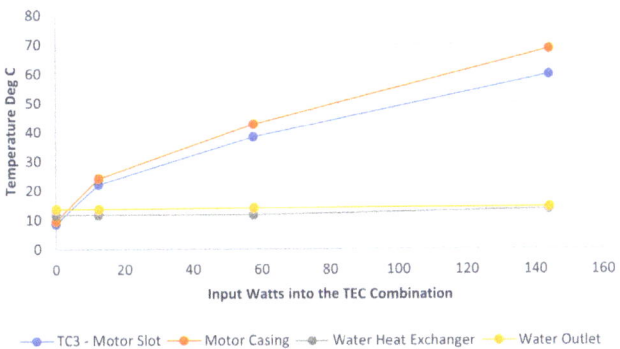

Figure 15. The graphed results of the heating with the TECs and water cooled heat sinks.

At 50 W nominal input power, the inner core and winding slot temperatures for all the heating systems are summarised in Table 1. In addition, the slot temperatures at various input powers for all the heating systems are shown in Figure 16.

Table 1. Comparison of the heating setups.

Setup	Resistive Heating	Heat Sink	Heat Pipe	Water Cooled
Arrangement	Three heating elements	14 TECs with heatsinks	Three TECs with heat pipes and radiators	10 TECs with water heat exchangers
Arrangement Cost	AUD 15.00	AUD 78.68	AUD 144.78	AUD 59.65
Complexity	Simple	Slightly Difficult	Slightly more difficult	Difficult
Physical Size	Minimal	Small	Large but flexible	Large when including heat exchanger
Performance at 50 W	32.14 °C	34.95 °C	42.93 °C	35.48 °C
Issues	Limited effectiveness.	Adapter plates for mounting. Effectiveness limited by heat sink size.	Adapter plates for mounting. Large heat exchangers. Flexible mounting by utilisation of heat pipes.	Adapter plates for mounting. Extra maintenance. Larger install area.

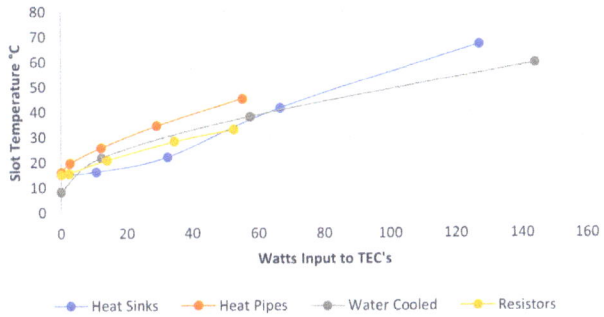

Figure 16. Comparison of the heating setup results.

4. Discussion

The resistive heating system, the most common system employed by manufacturers because of its low cost and ease of installation, is normally mounted internally. In this position, it heats the internal air, limiting its ability to heat the entire volume of the machine. There is thermal loss via the external surfaces, and the internal volume conducts the heat to these surfaces. This thermal path causes a thermal lag between the internal and external temperatures and limits the resistive heating systems' ability to stop internal condensation. The resistive system arrangement used in this study attached the resistive heating elements to the outside casing of a motor, and as this was the normal radiating surface, in this position, they could not be insulated, thus, the losses to the surroundings were high. The location had one major advantage; it was attached directly to the steel core, which conducted the heat quickly to the inner winding slot area. The results showed that at the nominal input power of 50 W, the winding temperature reached 32.14 °C, only a few degrees cooler than the other systems despite its simplicity. The results also showed that this system could make a difference, especially to very small motors where there is not enough room to internally mount resistive heaters.

In the TEC and heat sink system, there was a combination of seven adapter plates and 14 TECs with heat sinks. This system was simple to install with the adapter plates. This enabled the largest coverage of the external surface, reducing the surface losses; however, the very small heat sinks limited the TEC's ability to transfer or collect and pump the heat into the core. These worked well, and the slot temperature was easily able to be pushed up to 68 °C. At the nominal input power of 50 W, the inner core reached 34.95 °C. The main limitation was the very small size of the heat sinks. Larger units would have improved the TEC's ability to draw heat from the heat sink; however, this inexpensive system worked well and proved the concept.

In the TEC and water circulation heat sink system, there were 10 TECs. This system was difficult to assemble and install, as water tubing with tube clamps was required to be fitted to each heat sink. In a water cooled system, there is a lot of extra equipment such as pumps, radiators and fans. This equipment significantly increased the required footprint for the installation and added considerably to the system costs and particularly the maintenance costs. The results were comparable with the heat sink system. The cold water holding the heat exchangers at 12 °C limited the ability of the TECs to pump heat into the motor, but at the nominal input power of 50 W, the inner core reached 35.48 °C.

From Table 1, the three TECs connected to the fan cooled radiators via heat pipes resulted in the best result compared to the other heating systems. With a nominal input power of 50 W, the inner core reached 42.95 °C. This was achieved with only a small area covered by three TECs. The advantage of this configuration was that it connected the TECs via the casing to the core, which efficiently conducted the heat to the inner core area, where the stator winding was located. This system also demonstrated the potential to locate the heat exchangers away from the motor. In confined conditions such as electric cars and portable compressor stations, this would enable the heat exchangers to be in a more appropriate area than the motor location. The disadvantage, however, is that the fan cooled radiators need to be located in an open area and are fairly large.

The three TEC heating systems employed in this paper are commonly found in thermal control applications. The mounting of these systems to the outer surface or flat surfaces is simple as the TECs are flat units. For round motors, a simple adapter plate can be used. For ribbed motors, a more complex adapter plate is required. These systems can be used post manufacture.

The TEC systems are inexpensive and have a major advantage over the existing heating systems in that they can pump heat directly into the volume of the motor, raising the temperature well beyond the dew point and eliminating the condensation issues. For new machines, the incorporation of TECs could be implemented during the design phase and could be placed between the core and casing.

5. Conclusions

This work has demonstrated that the application of TECs can effectively pump heat into the core of electrical motors and generators, raising the core temperatures. All three TEC heat pumping systems were effective, and the choice between the approaches would be application driven. The results of applying the TECs in these three cases showed that for all three arrangements, the application of 50 W of power to the TECs resulted in an increase in the slot temperature of a 500 W induction motor to more than 20 °C above the ambient temperature.

TECs can pump heat in either direction, and previous studies have shown that TECs can pump heat out of a motor. A future study will examine the viability of dynamically managing a stable electrical motor core temperature in a varying ambient environment by utilizing controlled TECs to pump heat into and out of the core as required.

Author Contributions: Conceptualization, S.L., R.M. and J.C.; Investigation, S.L.; Project administration, R.M.; Resources, M.L. and J.C.; Supervision, R.M. and M.L.; Writing—original draft preparation, S.L. and R.M. Writing—review and editing, S.L., T.O., J.C., M.L. and R.M. All authors have read and agreed to the published version of the manuscript.

Funding: This research received no external funding.

Data Availability Statement: The data presented in this study are available on request from the corresponding author.

Conflicts of Interest: The authors declare no conflict of interest.

References

1. Maughan, C.; Gibbs, E.; Giaquinto, E. Mechanical testing of high voltage stator insulation systems. *IEEE Trans. Power Appar. Syst.* **1970**, *PAS-89*, 1946–1954.
2. Cowern, E. Keep Your Motor Dry in Damp Environments. 2001. Available online: https://www.ecmweb.com/content/article/20886991/keep-your-motor-dry-in-damp-environmrnts (accessed on 19 September 2012).
3. Kim, C.; Lee, K. Numerical Investigation of the Air-Gap Flow Heating Phenomena in Large-Capacity Induction Motors. *Int. J. Heat Mass Transf.* **2017**, *10*, 7. [CrossRef]
4. Jeftenic, I.; Stankovic, K.; Kartalovic, N.; Loncar, B. Life Expectancy Determination of Form-Wound Coil Isolation of High-Voltage Motor. In Proceedings of the 2016 IEEE International Power Modulator and High Voltage Conference (IPMHVC), San Francisco, CA, USA, 6–9 July 2016. . [CrossRef]
5. Nemirovskiy, A.; Kichigina, G.; Sergievskaya, I.; Udaratin, A.; Alyunov, A.; Gracheva, Y. Improving the Efficiency of Electroosmotic Drying of Electric Motors Insulation. *E3s Web Conf.* **2020**, *178*, 01061. [CrossRef]
6. Khudonogov, A.M.; Khudonogov, I.A.; Dulskiy, E.Y.; Ivanov, P.Y.; Lobytsin, I.O.; Khamnaeva, A.A. Reliability Analysis of Power Equipment of Traction Rolling Stock Within the Eastern Region. *IOP Conf. Ser. Mater. Sci. Eng.* **2020**, *760*, 012018. [CrossRef]
7. Zhang, P.; Du, Y.; Habetler, T.G.; Lu, B. A Nonintrusive Winding Heating Method for Induction Motor Using Soft Starter for Preventing Moisture Condensation. *IEEE Trans. Ind. Appl.* **2011**, *48*, 117–123. [CrossRef]
8. Keithly, W.R.; Axe, S.P. A unique solution to improving motor winding life in medium-voltage motors. *IEEE Trans. Ind. Appl.* **1984**, *IA-20*, 514–518.
9. Li, W.; Cotton, I.; Lowndes, R. Effects of Electrical Conductivity of Contamination on Tracking Formation in Aerospace Electrical Systems. In *Lecture Notes in Electrical Engineering, Proceedings of the 21st International Symposium on High Voltage Engineering, Budapest, Hungary, 26–30 August 2019*; Németh, B., Ed.; Springer: Cham, Switzerland, 2020; Volume 599. [CrossRef]
10. EASA. EASA Technical Manual. 2002. Available online: https://easa.com/resources/easa-technical-manual (accessed on 19 September 2012).
11. Gallonia, E.; Parisia, P.; Marignettib, F.; Volpec, G. CFD Analyses of a Radial Fan for Electric Motor Cooling. *Therm. Sci. Eng. Prog.* **2018**, *8*, 470–476. [CrossRef]
12. Giangrande, P.; Madonna, V.; Nuzzo, S.; Galea, M. Moving Toward a Reliability-Oriented Design Approach of Low-Voltage Electrical Machines by Including Insulation Thermal Aging Considerations. *IEEE Trans. Transp. Electrif.* **2020**, *6*, 16–27. [CrossRef]
13. Pai, V.S. Preservation of large motors and generators from weather on offshore platforms. *IEEE Trans. Ind. Appl.* **1990**, *26*, 914–918. [CrossRef]
14. Guastavino, F.; Torello, E.; Dardano, A.; Bono, S.; Pellegrini, M.; Vignati, L. A comparison of the short and long term behaviour of nanostructured enamels with the performances of conventi. In Proceedings of the Seventh International Conference on Machine Learning and Cybernetics, Kunming, China, 12–15 July 2008.
15. Branch, F.E. Keeping Motor Windings Dry. 1991. Available online: https://www.usbr.gov/power/data/fist/fist3_4/vol3-4.pdf (accessed on 10 September 2022).

16. Thomas, R. Preventing Condensation in 3-Phase AC Motors. 1996. Available online: https://www.ecmweb.com/content/article/20888577/preventing-condensation-in-3phase-ac-motors (accessed on 19 September 2012).
17. Finley, W.; Wilson, C.; Burke, R. Storage of electric motors. Conference Record of 1995. In Proceedings of the Annual Pulp and Paper Industry Technical Conference, Vancouver, BC, Canada, 6 August 1995; pp. 74–81. . [CrossRef]
18. Lysenko, D.A.; Konyukhov, V.Y.; Astashkov, N.P. Implementation of the control algorithm of the traction electric equipment. *J. Phys. Conf. Ser.* **2021**, *2061*, 012135. [CrossRef]
19. Livotov, P.; Sekaran, A.P.C.; Mas'udah; Law, R.; Reay, D. Eco-Innovation in Process Engineering—Contradictions, Inventive Principles and Methods. *Therm. Sci. Eng. Prog.* **2019**, *9*, 14.
20. U.S. Department of Energy. Chapter 6 Innovating Clean Energy Technologies in Advanced Manufacturing Direct Thermal Energy Con 2015, p. 26. Available online: https://www.energy.gov/sites/default/files/2015/12/f27/QTR2015-6G-Direct-Thermal-Energy-Conversion-Materials-Devices-and-Systems.pdf (accessed on 19 September 2012).
21. Sunawar, A.; Garniwa, I.; Hudaya, C. The characteristics of heat inside a parked car as energy source for thermoelectric generators. *Int. J. Energy Environ. Eng.* **2019**, *10*, 347–356. [CrossRef]
22. Lucas, S.; Marian, R.; Lucas, M.; Ogunwa, T.; Chahl, J. Research in life extension of electrical motors by controlling the impact of the environment through employing Peltier effect. *Energies* **2022**, *15*, 7659. [CrossRef]
23. Hebei, I.T. TEC1-12706 Thermoelectric Cooler. Shanghai Co Ltd. Available online: https://www.hebeiltd.com.cn (accessed on 20 November 2022).
24. Espressomilkcooler. Specification of Thermoelectric Module TEC1-12706. 1997. Available online: https://espressomilkcooler.com/wp-content/uploads/2015/03/TEC1-12706-site-ready.pdf (accessed on 20 November 2022).
25. Adera, S.; Antao, D.; Raj, R.; Wang, E.N. Design of Micropillar Wicks for Thin-Film Evaporation. *Int. J. Heat Mass Transf.* **2016**, *101*, 15. [CrossRef]
26. Bai, C.; Qiu, Y.; Wei, M.; Tian, M. Converging-Shaped Small Channel for Condensation Heat Transfer Enhancement Under Varying-Gravity Conditions. *Int. Commun. Heat Mass Transf.* **2021**, *123*, 11. [CrossRef]

Disclaimer/Publisher's Note: The statements, opinions and data contained in all publications are solely those of the individual author(s) and contributor(s) and not of MDPI and/or the editor(s). MDPI and/or the editor(s) disclaim responsibility for any injury to people or property resulting from any ideas, methods, instructions or products referred to in the content.

Article

Design Modifications for a Thermoelectric Distiller with Feedback Control

Mohammad Tariq Nasir [1,*], Diaa Afaneh [2] and Salah Abdallah [1]

[1] The Mechanical and Industrial Engineering Department, Applied Science Private University, Amman 11931, Jordan
[2] Mechanical Engineering Department, King Fahd University for Petroleum and Minerals, Dhahran 31261, Saudi Arabia
* Correspondence: mo_nasir@asu.edu.jo

Abstract: In this paper, a modified design for a thermoelectric distiller is proposed, constructed, and tested. The design modifications include adding an inclined cover for the thermoelectric module and a cooling fan. The thermoelectric module and the fan were operated by an open loop or a feedback control to have the desired productivity. As the distiller productivity depends on the operating conditions, these operating conditions are investigated to find the best performance with the highest pure water productivity. Furthermore. a comparison between the closed and the open loop for driving the cooling fan with different operating conditions is conducted. In this work, the mathematical model of the proposed distiller is derived. Experimental results illustrate the robustness of the proposed approach and they show that the suggested thermoelectric distiller with feedback control, for both cases with MPC and PID controllers, can increase pure water productivity by up to 150% when compared with the open loop thermoelectric distiller.

Keywords: water distillation; thermoelectric module; feedback; MPC controller; PID controller; system identification

Citation: Nasir, M.T.; Afaneh, D.; Abdallah, S. Design Modifications for a Thermoelectric Distiller with Feedback Control. *Energies* **2022**, *15*, 9612. https://doi.org/10.3390/en15249612

Academic Editor: Diana Enescu

Received: 5 November 2022
Accepted: 15 December 2022
Published: 18 December 2022

Publisher's Note: MDPI stays neutral with regard to jurisdictional claims in published maps and institutional affiliations.

Copyright: © 2022 by the authors. Licensee MDPI, Basel, Switzerland. This article is an open access article distributed under the terms and conditions of the Creative Commons Attribution (CC BY) license (https://creativecommons.org/licenses/by/4.0/).

1. Introduction

Distilled water is essential in many human practices, as it is used for many industrial and medical applications, such as the manufacture of medicines, laboratory analyzes, the manufacture of batteries, and ironing clothes [1]. However, the process of producing distilled water is challenged by the high energy consumption, due to the need to evaporate the water before condensing it [2]. It should be noted that the latent energy and the heat capacity of water are significantly high, leading to the consumption of a lot of energy, most of which is wasted. This high energy consumption is reflected in the environment since most distillation devices depend on electrical energy generated mostly from burning fossil fuels. Therefore, the active research community has studied several methods to obtain energy-efficient, low-cost, environment-friendly, and high-yield distillers.

Recently, many environment-friendly distillation designs tackled the challenge of adding a thermoelectric module (TEM) within the distiller apparatuses. TEM is a semiconductor device, that works based on the Seebeck effect. Thus, TEM is used in two ways. The first TEM application includes converting electrical energy to heat flux. The second usage opposes the first one because the TEM is exposed to heat flux converted into electrical energy. Recently, there has been considerable interest in TEM scientific research, in particular for the environment-friendly distillation process, and this is due to TEM's interesting properties. One of its remarkable properties is its ability to heat and cool simultaneously. In other words, TEM can generate heat flux or pump heat energy between its hot and cold sides. This process results in temperature differences between TEM sides. Because of this property, TEM can be utilized for water distillers, mainly considering three methodologies: (1) TEM is used as a secondary element for increasing the water condensation rate by placing the TEM cold side in touch with the vapor. Several works considered using the thermoelectric module to enhance the vapor condensation in solar distillers [3–6]. (2) TEM is a secondary element to increase water evaporation in solar distillers. In this case, the TEM hot side is attached to the saline water tank or the solar distiller basin [7]. (3) TEM is a primary source for heating the saline water and cooling the vapor. In this case, the TEM hot side is attached to the hot tank while the other TEM side is placed on the distiller cold tank. This way, the TEM restores the heat from the vapor and pumps it back to the hot tank for heating and boiling the saline water. Recently, several research methodologies were proposed considering TEM as a primary element. For example, the multi-stage design [8], the concentric design of the thermoelectric distiller [9,10], and the usage of thermoelectric distiller in urine–water recovery systems inside spacecraft [10,11]. Additionally, recent works tackled the use of the thermoelectric module to enhance vapor condensation and cooling in solar distillers [3,12].

Another interesting study presented thermoelectric utilization to improve saline water evaporation and condensation [13,14]. Such a proposed device employed a pump for recirculating the saline water between the water basin and the heat exchanger coupled to the hot side of the thermoelectric module. The cold side of the thermoelectric module is used for vapor condensation. This approach shows better energy conservation. However, the limitation of using a pump for water circulation will increase the required operational power. Moreover, to the best of our knowledge, no control strategy is proposed resulting in optimal productivity. A promising work proposed a simplified mathematical model for a thermoelectric-based distiller and compared it through experimental validation [15]; it proposed a new distiller design that is very efficient and has lower power consumption with high-purity water production. Nonetheless, the shortcoming of the work by [15] is wasting 45% of the input water through the system vent. This considerable loss is due to the vapor production rate being higher than the condensation rate. As a result, a considerable amount of vapor comes out of the distiller through a vent to conserve the constant pressure in the system. Similar work proposes a thermoelectric-based distiller by deriving a polynomial function for predicting the water evaporation rate with time [16]. The second TEM property is the ability to generate electrical power through the temperature difference between hot and cold sides. This property is employed to generate electricity from the heat losses in distillers [17]. The main challenge with this type of TEM application is its low efficiency since the thermoelectric generator efficiency is less than 20% [18]. This limitation reduces the TEM usage as a generator for low-temperature harvesting applications (e.g., water distillers).

Motivated by the abovementioned challenges, in this work, a modified TEM distiller is proposed, where the TEM is considered as a primary element such that both hot and cold sides are utilized in the evaporation and condensation stages. The main objective of this work is to improve the productivity of the TEM distiller through different design modifications summarized as follows:

1. The main contribution is the modified thermoelectric design. This new design includes adding a heat sink with a fan to the distiller cold tank outer face to attenuate the heat effect from the system. This facilitates controlling the temperature of the distiller cold tank.
2. Proposing a closed loop feedback control method where the output performance is compared with standard open loop control methods.
3. Proposing and implementing two feedback control systems, PID and MPC controllers, to control the operation of the thermoelectric distiller by manipulating the rotational speed of the fan and the TEM input voltage in the MPC case. The proposed control strategy successfully handles various unknown disturbances keeping the system at the desired operating conditions.
4. Additionally, an inclined cover of the thermoelectric distiller has been introduced to facilitate the collection of condensed drops of distilled water.

2. System Description

In this section, the proposed system parts and working principles are illustrated. Main distiller design modifications will be presented, followed by the mathematical model of the system, which consists mainly of energy and heat balance equations. Finally both steady-state and dynamic thermal model is derived.

2.1. The Proposed Thermoelectric Distiller Design

The proposed design of the thermoelectric module distiller (TEMD) uses the hot side of the thermoelectric module (TEM) to heat the water to the boiling point. The cold side is used to condensate the vapor and to recover a part of the vapor's latent heat to the hot side for reusing it. Figure 1 shows the main parts of the TEMD, which are two separate (casted aluminum) tanks, hot and cold tanks (9 × 5 × 5 cm each tank), a cover, and two TEM modules. The right tank includes hot water (saline water), and the left tank is for distilled water. The two TEMs are sandwiched between these two tanks in parallel. The condensed water in the cold tank loses a part of its latent heat to be pumped back to the hot tank by TEM. As a result, the waste energy will be reduced. However, heat losses occur in the hot and cold tanks to the surroundings, and vapor escapes from the vent when the distiller pressure increases. So, to keep the system in steady, a heat sink with a cooling fan is used to increase cooling and prevent water vapor from accumulating inside the system by condensing it and avoiding any vapor losses. The heat sink is fixed on the top of the cover over the cold tank. The fan is mounted above the heat sink. The primary function of the fan and the heat sink is to get rid of the excessive amount of energy to keep the process in a balanced state with maximum productivity.

The saline water tank is piped with the hot tank from both bottoms. This connection decreases heat loss because hot water has a lower density than cold water. As a result, the hot water will not flow through the pipe, minimizing heat losses. On the other hand, controlling the water level in the hot tank will be easily obtained by controlling the water level in the saline water tank. The water level is very low in the cold tank, and the condensed water film will cover the inside walls. The cold tank is attached to a U-shaped pipe to prevent the vapor from escaping from the tank while allowing only the distilled water to leave. A vent through the cover of the distiller is connected with a long pipe. This arrangement is required to keep the inside pressure near the atmospheric pressure and to prevent vapor accumulation in the hot tank.

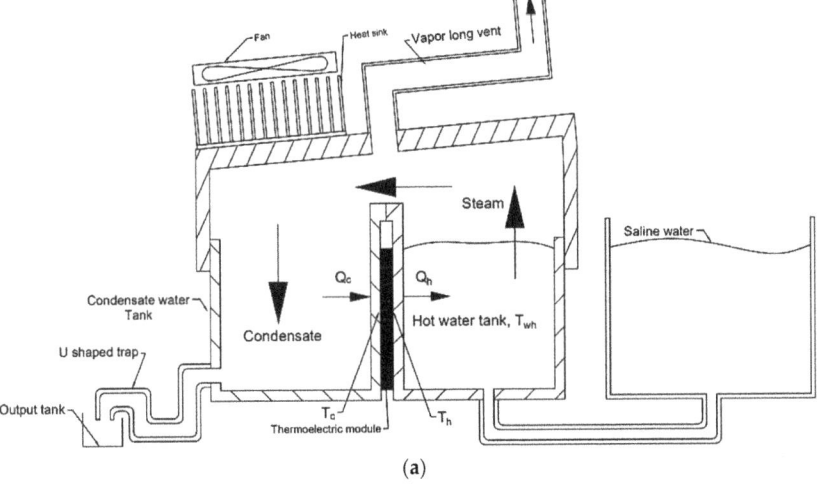

(a)

Figure 1. *Cont.*

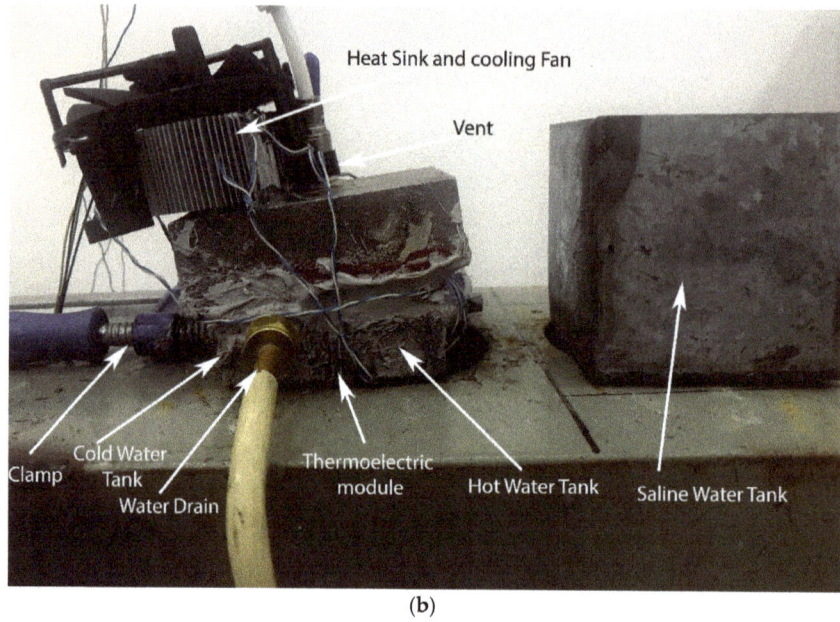

Figure 1. (**a**) TEMD schematic diagram. (**b**) TEMD experimental setup.

The inside cover face is inclined to make the condensate water drops slide to the cold tank, which increases productivity.

2.2. Mathematical Modeling of the System

In this section, a mathematical model of the proposed system is derived. A similar model was validated in [15]. This model represents energy balance, mass balance, and heat transfer equations. The thermal model can be simplified into three main areas: the TEM model, the hot tank model, and the cold tank model, as shown in Figure 2.

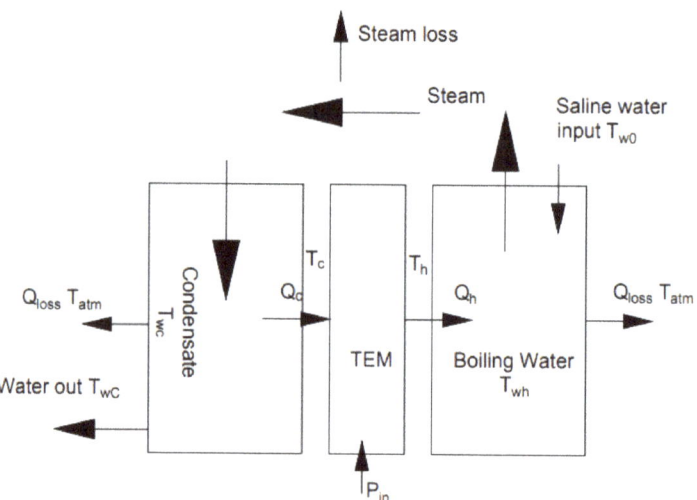

Figure 2. The thermal model diagram of TEMD.

The first thermal zone is TEM. The TEM model is given in [18,19]. Since we have an n number of the TEMs connected thermally in parallel, and electrically in series, the equations will be as follows:

$$Q_h = n\left[\alpha I T_h - k\Delta T + \frac{1}{2}I^2 R\right] \quad (1)$$

$$Q_c = n\left[\alpha I T_c - k\Delta T - \frac{1}{2}I^2 R\right] \quad (2)$$

$$V = \frac{IR}{n} + \alpha \Delta T \quad (3)$$

where Q_h and Q_c are, respectively, the heat flow rate to the hot side and the heat flow rate entering from the cold side of the thermoelectric module. α is the Seebeck coefficient, R is the electric resistance, k is the thermal conductance of the thermoelectric module, and $\Delta T = T_h - T_c$. The thermoelectric module type used is TEC1-19908, and its characteristics are validated experimentally by applying a simple least square fitting procedure for collected voltage, current, and temperature readings as in Equation (3). The TEM characteristics are $\alpha \cong 0.088$ V/K, $R \cong 2.38\ \Omega$, and $k \cong 0.8889$ W/K.

The second thermal zone is for the hot tank. Heat convection from the hot side of TEM to the water inside the hot tank is found using Equation (4), and heat loss from the water in the hot tank to the surroundings is found in Equation (5):

$$Q_h = \frac{T_h - T_{wh}}{R_h} \quad (4)$$

$$Q_{H,loss} = \frac{T_{wh} - T_{amp}}{R_{out}} \quad (5)$$

The resultant dynamic energy balance equation is:

$$Q_h - Q_{H,loss} - \dot{m}_{v,loss}h_v + \dot{m}_{win}h_{win} - \dot{m}_v h_v = \frac{dT_{wh}}{dt}V_{wh}\rho_{wh}C_p \quad (6)$$

The related steady energy balance equation is:

$$Q_h - Q_{H,loss} - \dot{m}_{v,loss}h_v + \dot{m}_{win}h_{win} - \dot{m}_v h_v = 0 \quad (7)$$

The vapor loss due to the vent and the corresponding vapor energy loss are given in the following equation:

$$\dot{m}_{v,loss} = \dot{m}_v - \dot{m}_{cond} \quad (8)$$

The third thermal zone is the cold tank volume. Equation (9) represents the convection heat transfer from the cold side of TEM to the condensed water film that covers the cold tank walls. Equations (10) and (11) formulate heat loss from the vapor and the condensed water film to the surroundings. For simplification, the thermal resistance is assumed to be the same for both losses.

$$Q_c = \frac{T_{wc} - T_c}{R_c} \quad (9)$$

$$Q_{C,loss} = \frac{T_{wc} - T_{amp}}{R_{out}} \quad (10)$$

$$Q_{V,loss} = \frac{T_v - T_{amp}}{R_{out}} \quad (11)$$

Equation (12) represents the dynamic heat balance in the cold tank, which can be used to find the condensation mass flow rate.

$$\dot{m}_{cond}h_v - Q_c - Q_{C,loss} - Q_{V,loss} - \dot{m}_{cond}h_{cond} = \frac{dT_{wc}}{dt}V_c\rho_c C_p \quad (12)$$

Equation (13) represents the steady-state heat balance.

$$\dot{m}_{cond}h_v - Q_c - Q_{C,loss} - Q_{Vloss} - \dot{m}_{cond}h_{cond} = 0 \qquad (13)$$

Note that the R_{out} in the cold tank depends on the fan rotational speed.

(A) To simplify the model, the following assumptions were used:
(B) $T_v = 98\ °C$ while boiling.
(C) $\dot{m}_v = \dot{m}_{win}$ in the hot tank.
(D) The distiller pressure is equal to the atmospheric pressure due to the small vent in the design.
(E) The thermal heat transfer resistances R_h, R_c, and R_{out} are constants.

Vapor loss can be calculated by $\dot{m}_{v,losses} = \dot{m}_v - \dot{m}_{cond}$, and it is found experimentally, where $\dot{m}_{v,losses} = 5\%\ \dot{m}_v$.

The system coefficient of performance can be calculated using the following equations:

$$COP_h = \frac{Q_h}{P_{in}} = COP_c + 1 \qquad (14)$$

$$COP_c = \frac{Q_c}{P_{in}} \qquad (15)$$

$$P_{in} = I * V \qquad (16)$$

The distiller mathematical model shows a strong relation between the fan and TEM's input currents and the system states. The following section discusses developing a control system that manipulates the voltage inputs for both the fan and TEM for obtaining better productivity.

3. Control System Description and the Model Identification

In this section, different control methodologies are proposed. Some of the methodologies are not suitable with the proposed system and others are too complicated. Then, the usage of two controller types are discussed.

3.1. The Power Circuit of TEMD System

The TEMD hardware consists of the hot tank, cold tank, saline water tank, and cover. All of these are cast aluminum, except the saline tank, which is made from a 1.0 mm galvanized steel sheet. Aluminum has high thermal conductivity and can help to cool the cold tank. On the other side, insulation means are needed on the hot tank to reduce energy losses. Figure 3 shows the schematic diagram of the TEMD as an electrothermal system. The electrical components are 24 VDC power supply generated by a photovoltaic system, a 12 buck converter, 2 PWM drivers, two TEMs, and two thermocouples with a MAX 6675 interface. The Arduino Mega is used to control the process, and it is connected to a PC and programmed with Simulink and MATLAB. The data are collected by using Simulink software. The TEM input voltage is set constant at 14 VDC with the PID controller, and the MPC controller controls its value, as will be discussed in the next section. The fan voltage is the controller command in both PID and MPC controllers. The fan is used to cool the cold side of the TEMD.

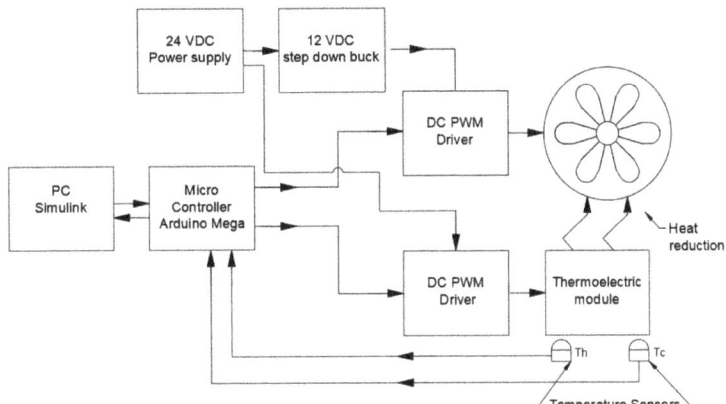

Figure 3. Control schematic diagram of TEMD.

3.2. Control Methodology

Previous work has tackled driving the thermoelectric distiller in open loop control [15]. This paper proposes two principal control methodologies: open loop and closed loop control. With an open loop, two cases are conducted, the first case is running the thermoelectric distiller without a cooling fan, and this case is like the previous work done in [15]. This open loop case showed that the vapor production rate is higher than the condensation rate; this main drawback led to increased vapor pressure in the distiller. The second open loop case has the cooling fan at full speed by setting the highest input voltage to the fan motor. In both open loop systems, the TEM's voltage is set constant such that the PWM voltage source is set to 14 VDC. In the results section, these open loop cases will be compared with proposed closed loop methodologies.

The second principal methodology is a closed loop system. This paper proposes closed loop control to drive the thermoelectric based water distillation process and guarantee maximum productivity. This target can be achieved by keeping the system balanced by having the water evaporation rate equal to the water condensation rate, so that the vapor loss will be minimized, and, as a result, the lost energy will be minimized. Since the vapor production rate is more than the condensation rate, an external cooling device is needed to keep the system balanced. This work adds a fan and heat sink to the TEMD system. The fan will run to minimize the extra vapor produced and keep the system balanced at a high heat recovery state. It was realized that when the fan runs at full rotational speed, more heat will be removed from the system, which will decrease the recovered energy by TEM to the hot tank. At this point, using a fan at full rotational speed will drop the system efficiency. On the other hand, if the fan is stopped, more vapors will be accumulated, and the vapor will escape from the system vent having more energy loss (the open loop case). So, it should be an optimum operating point in the middle with a certain value of the fan speed with maximum productivity and minimized energy losses.

Three proposed closed loop strategies may possibly be used for obtaining maximum productivity, as follows:

1. The first one is keeping a constant voltage for the TEM and controlling the fan rotational speed based on the pressure inside the system with no vapor vent. So, if the vapor production increases, the pressure will rise accordingly, and the controller will increase the fan speed to reduce the vapor and the pressure. This method was experimentally found to be difficult since the distiller vapor pressure value varies rapidly, and it requires to keep the system working at the equilibrium point by controlling the system's input and output water flows. This control approach is similar to the steam power plant control system. Therefore, this method is not considered in this research.

2. The second methodology considers the TEM cold side temperature T_c as a feedback signal and the fan input voltage as a control command, while keeping the vent open with constant voltage at the TEM input. The controller's response is simplified into two steps: 1—TEM cold side temperature increases, and more vapor will be generated in the TEMD. 2—Then the controller will increase the fan speed to condense more vapor and keep the system at a stable T_c operating point, sustaining lower pressure and decreasing the vapor losses. Experimentally the TEM cold side temperature indicates the amount of vapor inside the system. However, the energy harvested from the vapor's latent heat will decrease if the fan's rotational speed is increased. A PID controller is implemented with this strategy and compared with the third one. The optimum value of the fan's rotational speed will be discussed in the results section.

3. The third control methodology considers more input signals T_c and T_h, compared with the second strategy, which uses only T_c. The third strategy deals with two control commands: the fan and TEM voltages. So, the system will be a multi-input and multi-output system. The controller will drive the system using a well-known control method called Model Predictive Control (MPC). Using a predefined mathematical model, this method predicts the system's behavior with the control commands. So, the controller will generate the optimum control commands that perform the best dynamic responses from studying the system's mathematical model. This third method is implemented and compared with the second control methodology.

3.3. PID and MPC Control System Schematic Diagrams

Figure 4 shows the schematic diagram of the suggested control system with a PID Controller. The PID controller is a common and a famous controller type, where the error signal e is derived as in Equation (17); then, the control command u, current or voltage in our case, is calculated using Equation (18). Three terms are used in this controller (proportional, integral, and derivative terms), and three parameters (k_p, k_i, and k_d) have to be tuned appropriately. PID controller tuning can be implemented by software once the system's mathematical model is given. The system model is found experimentally, and will be discussed in the following subsection. In the PID controller case, the distiller has two inputs: the TEM input voltage and the fan input voltage. The TEM voltage is set to be a constant. The PID controller manipulates the fan voltage to keep the T_c at the desired value T_{c_d}.

$$e = T_{c_d} - T_c \tag{17}$$

$$u = e * k_p + k_i * \int e\, dt + k_d \frac{de}{dt} \tag{18}$$

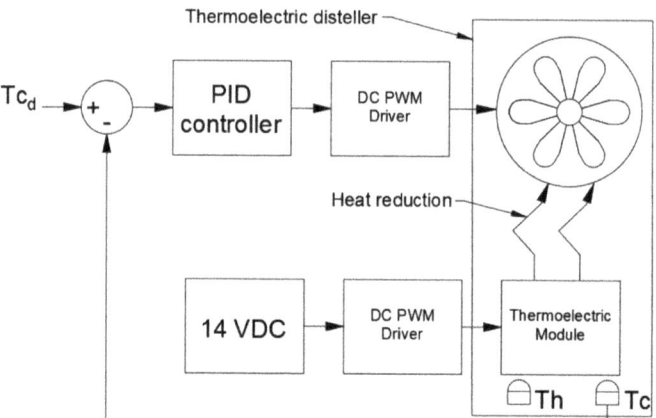

Figure 4. Schematic diagram of the system with PID Controller.

Figure 5 shows the schematic diagram of the suggested control system with the MPC controller. In the case of the MPC controller, the distiller has two inputs: the TEM input voltage and the fan input voltage. Temperature sensors send the feedback to the controller, and both of them are manipulated by the controller. The MPC controller is an optimal controller type [19]. MPC predicts the system dynamics for a finite length of time, called a prediction horizon. The system model is required and found experimentally, as will be discussed in the following subsection. Using the control command u, MPC minimizes a quadratic cost function J as shown in Equation (19).

$$J = \sum_{i=1}^{n} w_{xi}(r_i - x_i)^2 + \sum_{i=1}^{n} w_{ui} u_i^2 \qquad (19)$$

where w_{xi} is the system state weight.

r_i are the desired values, in our case T_{c_d} and T_{h_d}.
x_i are the system states, in our case T_c and T_h.
w_{ui} are the weights of the control commands.
u_i are the system control commands, in our case the TEM and fan voltages.

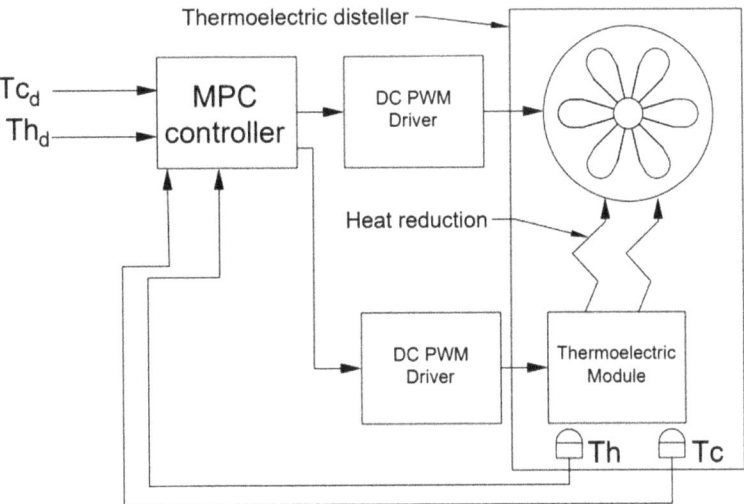

Figure 5. The schematic diagram of the system with MPC controller.

3.4. System Identification: MPC and PID Controller Tuning

A properly identified TEMD model is crucial for tuning the PID and MPC controllers. TEMD model is complicated since the heat transfer coefficients are functions of steam pressure inside the system. One easy method to simplify this model is using MATLAB identification tools that give an approximated model around the operating points. The identification method starts with running the model in open loop and collecting the output values accordingly during a specific time interval. Since the TEMD system has two inputs (TEM voltage and fan voltage) and two outputs (T_h and T_c), this control system can be considered a multi-input multi-output (MIMO) system. The curves of input voltages and output temperatures of this system are shown in Figure 6.

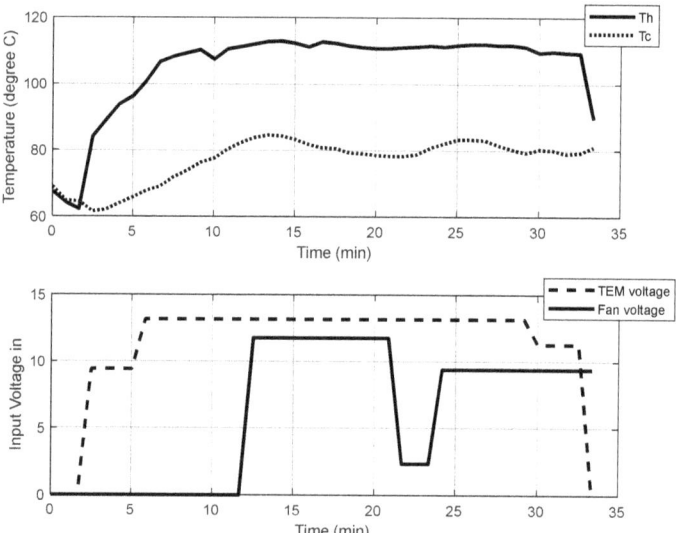

Figure 6. The open loop system temperature responses based on the manually adjusted input voltages.

It is noted that, at the beginning when switching the system on, the T_c will decrease slightly due to the Seebeck effect. Then after a short time, T_h will increase, and the T_c will increase accordingly while keeping a temperature difference.

These results are used in the identification toolbox in MATLAB with command N4SID to have the best approximated discrete state-space model in Equations (20) and (21). The identified model was found with sample time equaling 1 s and A, B, C, and D matrices as shown below. The identified model is a 95% fit with a final prediction error (FPE) of 0.01427 and a mean squared Error (MSE) of 0.2683. These values indicate that this model is accepted.

$$x(t + T_s) = Ax(t) + Bu(t) \tag{20}$$

$$y(t) = Cx(t) + Du(t) \tag{21}$$

where $x = [T_h, T_c]'$ and $u = [V_{TEM}, V_{fan}]'$.

$$A = \begin{bmatrix} 0.984 & 0.0153 \\ 0.0041 & 0.9958 \end{bmatrix}, B = \begin{bmatrix} 0.051 & -0.0107 \\ -0.0073 & -0.00264 \end{bmatrix}, C = \begin{bmatrix} 1 & 0 \\ 0 & 1 \end{bmatrix}, D = \begin{bmatrix} 0 & 0 \\ 0 & 0 \end{bmatrix}$$

The values of tuning parameters for PID controller are found to be $K_P = 4.05$, $K_I = 0.00024$, and $K_D = 1.0494$. The PID controller has a limited range of voltage equal [0–12] VDC with the clamping anti-wind up method. In addition, this model is fed into the MPC toolbox in MATLAB.

4. Experimental Results

This section discusses the results of the open loop and closed loop control strategies, suggests optimal operating conditions found experimentally, and compares the experimental results of the closed loop strategies (PID and MPC), including the system dynamics with both PID and MPC controllers.

4.1. Open Loop Experimental Results

Two open loop cases were discussed in previous sections; the first case is running the distiller without a cooling fan, and the experimental results show system productivity equal to 120 mL/h of purified water. The results of the second open loop case with maximum fan speed show 125 mL/h purified water productivity. Both cases have approximately low production rates and low productivity to power ratios, as shown in Table 1. The explanation for this decrease in productivity is due to water loss in the vent in case 1, and the increase of energy loss with the fan in case 2. For more details, refer to Section 3.2.

Table 1. Experimental results.

TEM PWM Voltage (Volt)	24 VDC Power Supply Current (Amp)	TEM Power (Watt)	TDS (ppm)	T_h (°C)	T_c (°C) (Set Point for Controlled Cases)	Productivity (ml/h)	Production to Power Ratio (ml/(W·h))	Fan Operation
14.2	5.78	138.72	54	110	67.5 *	125	0.9	Full power (open loop case 2)
14	5.81	139.44	62	111	72.5	157.9	1.13	
14	5.85	140.4	68	111	74	150	1.07	
14	5.84	140.16	70	111	74	166.7	1.19	by closed loop feedback controller
14	5.85	140.4	74	111.5	78	176.5	1.26	
14	5.8	139.2	75	112.3	82	187.5	1.35	
14	6.08	145.92	83	112	84	166.7	1.14	
14	5.97	143.28	115	111	84	187.5	1.31	
14	5.88	141.12	180	112	89 *	120	0.85	Off (open loop case 1)

* Without control (open loop).

4.2. Closed Loop Optimum Operating Conditions

The approach used in this paper is to find the optimal operating point experimentally and then compare the two proposed closed loop control strategies. Consequently, many experimental tests were performed to find the relation between the system productivity and operating conditions T_c and T_h. Table 1 shows the experimental results with different T_c set points. Compared to previous work [14], the productivity to power ratio is decreased because of the heat loss in the electronic devices and PWM drivers. That causes transistors to consume extra heat loss due to the high frequency switching with the PWM method.

The productivity to power ratio varies with respect to T_c as shown in Figure 7. The best production to power ratio is at set point $T_c = 80$ °C. T_h mainly depends on the hot water temperature in the hot tank, and since it is constant (equal to 98 °C), there are no major changes in T_h. The behavior of the system can be illustrated as follows: low T_c id caused by the high cooling from the fan, so the system will lose more energy and the evaporation rate will decrease. On the other hand, high T_c means low cooling from the fan; the vapor will accumulate due to the weakened condensation rate, and then vapor losses will increase via the vent, which will reduce productivity. As shown in Table 1, the optimum production to power ratio is at $T_c = 82$ °C, which is 150% more compared to open loop operating conditions: (1) fan switched off and (2) full fan power.

In summary, water productivity in the proposed operating conditions with closed loop control is higher than the value of the open loop thermoelectric distiller. This shows that the control of the TEM cold side temperature plays an important role in thermoelectric efficiency and productivity. The resultant specific energy consumption (SEC) at the max productivity is equal to 742 kWh/m^3. Note that the best operating conditions found experimentally are not fixed, and they may vary if the distiller design changes. The max productivity rate to power ratio is 1.35 mL/(W·h) compared with 0.49 mL/(W·h) by previous thermoelectric based distiller proposed in [13].

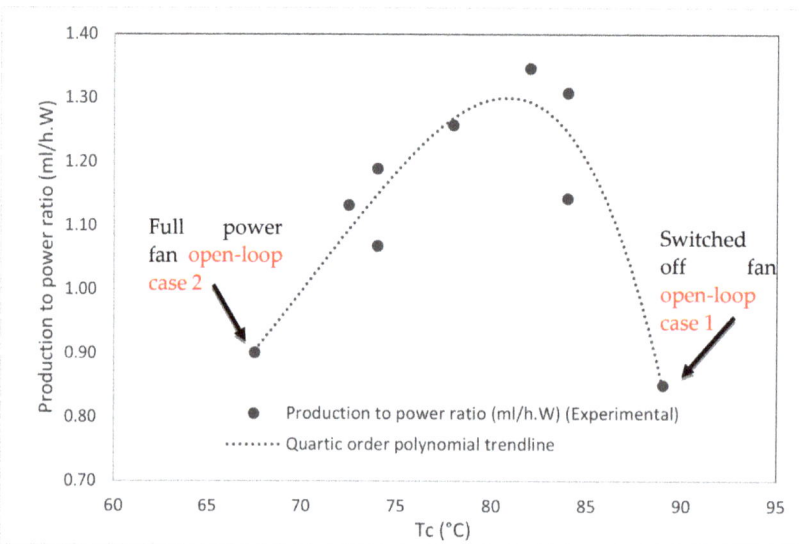

Figure 7. The system's production to power ratio with different T_c values and constant TEM voltage = 14 VDC.

4.3. Testing PID Controller

The PID controller is added to the system then the system is closed loop operated. Figure 8 shows the system response with the proposed PID controller, where the settling time is 20 min. This controller is used to regulate the system in the desired operating conditions, in this case keeping T_c at 80 °C.

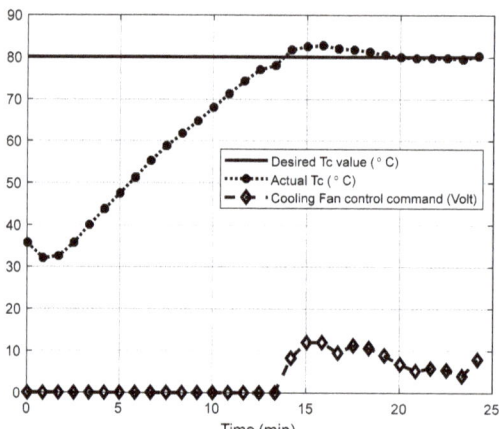

Figure 8. The system response with the proposed PID controller.

4.4. Comparing PID with MPC Controllers

The model predictive controller (MPC) is a multi-input and multi-output controller that can, in this case, send two control commands: the TEM input voltage and fan input voltage for regulating the two outputs, T_h and T_c values. This control approach is compared with the well-tuned PID controller, and the results were found as shown in Figure 9. The MPC shows faster system performance with less settling time. This result was expected since the MPC controller controls the TEM voltage and the fan speed, compared with only controlling the fan speed with PID controller.

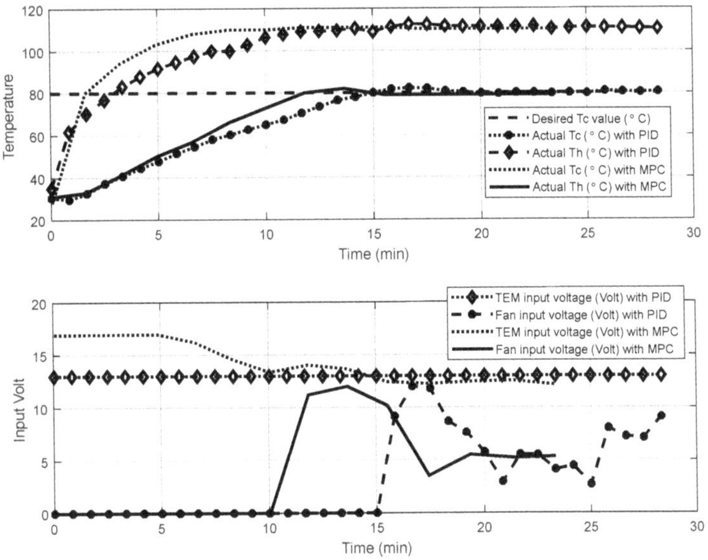

Figure 9. TEMD dynamic performance with the PID and MPC controllers.

The operating conditions of the modified design were determined experimentally by studying the effect of varying the cold and hot side temperatures on the system's productivity. Two control methodologies were proposed, PID and MPC, and both strategies were used to keep the system working at specified operating conditions. First, the system model around the operating point is determined experimentally, and then the PID and the MPC controllers are adjusted based on this model. The effect of the added design modifications on the thermoelectric distiller design is clear. The total collected distilled water during 20 min from the previous design thermoelectric distiller was 0.85 mL/(W·h) [13], and the total collected distilled water from the modified thermoelectric distiller was 1.35 mL/(W·h). The MPC controller leads to better dynamic performance than the PID controller.

5. Conclusions

The performance of a novel design of a thermoelectric distiller is investigated. The design modifications include the addition of an inclined cover of the thermoelectric module, a heat sink with a controlled cooling fan, using two feedback control systems with PID and MPC controllers, and the suggestion of a special control strategy to handle various disturbances and variables to keep the system in a steady state at the desired operating conditions. Different operating conditions are investigated to find the best performance with the highest pure water productivity. In this work, the mathematical model of the proposed distiller is derived. A comparison between PID and MPC controllers with different operating conditions is conducted. The system with the MPC controller has better dynamic performance than the system with PID controller. Experimentation results show that a thermoelectric module with the proposed design and operating conditions, for both cases with MPC and PID controllers, increase productivity by up to 150%, compared with the open loop thermoelectric distiller. This increase is obtained at the optimum operating conditions because of increasing the heat recovery and vapor productivity and decreasing the vapor loss in the vent. The results show higher productivity with less water losses compared with previous thermoelectric distillers.

Author Contributions: The study conception and design were performed by M.T.N., D.A. and S.A. contributed to the study conception and design. The controller design was performed by M.T.N., and data collection and analysis were performed by M.T.N. and D.A. The first draft of the manuscript was written by M.T.N. All authors have read and agreed to the published version of the manuscript.

Funding: The work is part of a funded project at Applied Science Private University with number DRGS-2021-2022-2.

Data Availability Statement: All data are provided in the manuscript in the form of tables or figures.

Acknowledgments: The authors are grateful to Applied Science Private University, Amman, Jordan for the full financial support granted to this research project.

Conflicts of Interest: The authors declare no conflict of interest.

Nomenclature

Symbol	Description
α	The Seebeck effect coefficient (V/K)
C_p	The heat capacity of the water (kJ/(kg·K))
COP_h	Coefficient of performance for heating process in the TEM
COP_c	Coefficient of performance for cooling process in the TEM
ΔT	$T_h - T_c$ (K)
h_v	Enthalpy of the vapor (kJ/kg)
h_{cond}	Enthalpy of the output water at 45 °C (kJ/kg)
h_{fg}	The latent heat of the boiled water (kJ/kg)
h_{win}	Enthalpy of the inlet water at 25 °C (kJ/kg)
I	The TEM inlet current. (A)
K_p	PID controller proportional gain
K_d	PID controller derivative gain
K_i	PID controller integral gain
k	Thermal conductivity of the TEM (W/K)
$\dot{m}_{win}$	The mass flow rate of the inlet water. (kg/s)
$\dot{m}_v$	The mass flow rate of the produced vapor (kg/s)
$\dot{m}_{cond}$	The mass flow rate of the produced water (condensation) (kg/s)
m_{hw}	The total water mass in the hot tank (kg)
$\dot{m}_{v,losses}$	Vapor loss flow rate outside the tanks from the vent (kg/s)
n	The number of TEM connected in parallel
$Power_{in}$	The electrical power consumed by the TEM $D = I * V$ (W)
P_{in}	Electrical power input to TEMs (W)
Q_c	Heat flux of the TEM cold side (W)
Q_h	Heat flux of the TEM hot side (W)
Q_{loss}	Heat losses from the hot or the cold tank to surroundings
R_c	Thermal resistance between the TEM cold side and the vapor (K/W)
R_h	Thermal resistance between the TEM hot side and the hot tank water (K/W)
R	The TEM electrical resistance (Ω)
R_{out}	The thermal resistance between the system and the environment (K/W)
ρ_{wh}	Hot water density
ρ_c	Condensate water density
T_h	TEM hot side temperature (K)
T_c	TEM cold side temperature (K)
T_{wh}	Temperature of the water in the hot tank (K)
T_v	Temperature of the vapor (K)
T_{wc}	Temperature of the water in the cold tank (K)
T_{atm}	Temperature of the surrounded environment (25 °C)
TDS	Total dissolved solids (ppm)
V	The voltage source value (V)
V_c	Condensate water drops covers inside of the cold tank wall.
V_{wh}	Hot water volume inside of the hot tank.

Abbreviations

Symbol	Description
A	Amber
°C	Degree Celsius
MPC	Model predictive controller
MIMO	Multi-input multi-output system
PID	Proportional integral derivative controller
PWM	Pulse-width modulation
ppm	Part per million
TEM	Thermoelectric module
TDS	Total dissolved solids
TEMD	Thermoelectric module-based distiller
VDC	DC voltage

References

1. Abou Assi, R.; Ng, T.F.; Tang, J.R.; Hassan, M.S.; Chan, S.Y. Statistical Analysis of Green Laboratory Practice Survey: Conservation on Non-Distilled Water from Distillation Process. *Water* **2021**, *13*, 2018. [CrossRef]
2. Al-Karaghouli, A.; Kazmerski, L.L. Energy Consumption and Water Production Cost of Conventional and Renewable-Energy-Powered Desalination Processes. *Renew. Sustain. Energy Rev.* **2013**, *24*, 343–356. [CrossRef]
3. Al-Nimr, M.A.; Al-Ammari, W.A. A Novel Hybrid and Interactive Solar System Consists of Stirling Engine vacuum Evaporator thermoelectric Cooler for Electricity Generation and Water Distillation. *Renew. Energy* **2020**, *153*, 1053–1066. [CrossRef]
4. Parsa, S.M.; Rahbar, A.; Koleini, M.H.; Aberoumand, S.; Afrand, M.; Amidpour, M. A Renewable Energy-Driven Thermoelectric-Utilized Solar Still with External Condenser Loaded by Silver/Nanofluid for Simultaneously Water Disinfection and Desalination. *Desalination* **2020**, *480*, 114354. [CrossRef]
5. Das, D.; Bordoloi, U.; Kalita, P.; Boehm, R.; Kamble, A. Solar Still Distillate Enhancement Techniques and Recent Developments. *Groundw. Sustain. Dev.* **2020**, *10*, 100360. [CrossRef]
6. Alwan, N.; Ahmed, A.; Majeed, M.; Shcheklein, S.; Yaqoob, S.; Nayyar, A.; Nam, Y.; Abouhawwash, M. Enhancement of the Evaporation and Condensation Processes of a Solar Still with an Ultrasound Cotton Tent and a Thermoelectric Cooling Chamber. *Electronics* **2022**, *11*, 284. [CrossRef]
7. Al-Nimr, M.A.; Qananba, K.S. A Solar Hybrid Thermoelectric Generator and Distillation System. *Int. J. Green Energy* **2018**, *15*, 473–488. [CrossRef]
8. Stout, B.; Peebles, R. High Temperature Peltier Effect Water Distiller. U.S. Patent 6,893,540, 17 May 2005.
9. Milton, R. Peltier Effect Concentric Still. U.S. Patent 3,393,130, 16 July 1968.
10. Trusch, R.B. Thermoelectric Integrated Membrane Evaporation System. U.S. Patent 4,316,774, 23 February 1982.
11. Samsonov, N.M.; Bobe, L.S.; Rifert, V.G.; Barabash, P.A.; Komolov, V.V.; Margulis, V.I.; Novikov, V.M.; Pinsky, B.Y.; Protasov, N.N.; Rakov, V.V.; et al. System and a Rotary Vacuum Distiller for Water Recovery from Aqueous Solutions, Preferably from Urine Aboard Spacecraft. U.S. Patent 6,258,215, 10 July 2001.
12. Shoeibi, S.; Rahbar, N.; Esfahlani, A.A.; Kargarsharifabad, H. Application of Simultaneous Thermoelectric Cooling and Heating to Improve the Performance of a Solar Still: An Experimental Study and Exergy Analysis. *Appl. Energy* **2020**, *263*, 114581. [CrossRef]
13. Al-Madhhachi, H. Solar Powered Thermoelectric Distillation System. Ph.D Thesis, Cardiff University, Cardiff, UK, 2017.
14. Al-Madhhachi, H.; Phillips, M.; Gao, M. *Validation of Vapour/Water Production in a Thermoelectric Distillation System*; Avestia Publishing: Ottawa, ON, Canada, 2017; p. 117-1.
15. Nasir, M.T.; Afaneh, D.; Abdallah, S. High Productivity Thermoelectric Based Distiller. *Desalination Water Treat.* **2020**, *206*, 125–132. [CrossRef]
16. Sasongko, S.B.; Sanyoto, G.J.; Buchori, L. Study of Performance: An Improved Distillation Using Thermoelectric Modules. *Chem. Eng. Trans.* **2021**, *89*, 649–654.
17. Khanmohammadi, S.; Chaghakaboodi, H.A.; Musharavati, F. A New Design of Solar Tower System Amplified with a Thermoelectric Unit to Produce Distilled Water and Power. *Appl. Therm. Eng.* **2021**, *197*, 117406. [CrossRef]
18. Snyder, G.J.; Ursell, T.S. Thermoelectric Efficiency and Compatibility. *Phys. Rev. Lett.* **2003**, *91*, 148301. [CrossRef] [PubMed]
19. Mayne, D.Q.; Rawlings, J.B.; Rao, C.V.; Scokaert, P.O. Constrained Model Predictive Control: Stability and Optimality. *Automatica* **2000**, *36*, 789–814. [CrossRef]

MDPI AG
Grosspeteranlage 5
4052 Basel
Switzerland
Tel.: +41 61 683 77 34

Energies Editorial Office
E-mail: energies@mdpi.com
www.mdpi.com/journal/energies

Disclaimer/Publisher's Note: The title and front matter of this reprint are at the discretion of the . The publisher is not responsible for their content or any associated concerns. The statements, opinions and data contained in all individual articles are solely those of the individual Editor and contributors and not of MDPI. MDPI disclaims responsibility for any injury to people or property resulting from any ideas, methods, instructions or products referred to in the content.

www.ingramcontent.com/pod-product-compliance
Lightning Source LLC
LaVergne TN
LVHW070713100526
838202LV00013B/1082